AI AWARENESS SERIES

AI in Insurance

Emilia Silverfield

Contents

Introduction 1

Chapter 1: Introduction to AI in Insurance 3

Chapter 2: Data Infrastructure and Readiness 15

Chapter 3: AI Techniques and Tools in Insurance 22

Chapter 4: AI-Driven Underwriting 30

Chapter 5: Risk Profiling and Scoring 41

Chapter 6: Fraud Detection and Prevention 50

Chapter 7: AI-Powered Claims Processing 60

Chapter 8: Enhancing Customer Experience 67

Chapter 9: Policy Servicing and Lifecycle Management 76

Chapter 10: Generative AI and New Product Development 85

Chapter 11: AI in Distribution and Sales 95

Chapter 12: Strategic AI Governance 102

Chapter 13: Health and Life Insurance 109

Chapter 14: Property and Casualty Insurance 117

Chapter 15: Commercial and Cyber Insurance 125

Introduction

The insurance industry stands at an unprecedented crossroads. After centuries of operating on relatively static models of risk assessment, policy management, and claims processing, artificial intelligence is fundamentally reshaping every aspect of how insurers operate, compete, and serve their customers.

This transformation extends far beyond simple automation. AI is revolutionizing the core functions that define insurance: from using machine learning algorithms to analyze vast datasets for more accurate underwriting, to deploying computer vision for instant damage assessment, to leveraging predictive analytics that can forecast risks before they materialize. What once required weeks of manual processing can now be accomplished in minutes with greater accuracy and consistency.

For insurance professionals, understanding AI is no longer optional—it's essential for survival in an increasingly competitive and technology-driven marketplace. Companies that successfully integrate AI capabilities are seeing dramatic improvements in operational efficiency, customer satisfaction, and financial performance. Those that resist this technological evolution risk being left behind as competitors leverage AI to offer faster service, more personalized products, and more competitive pricing.

This book provides a comprehensive roadmap for navigating the AI revolution in insurance. From foundational concepts in data

Introduction

infrastructure and machine learning techniques to advanced applications in fraud detection, customer experience enhancement, and emerging areas like cyber insurance, each chapter builds practical understanding that can be immediately applied to real-world insurance challenges.

Whether you're an insurance executive charting digital transformation strategy, an underwriter seeking to understand AI-powered risk assessment, a claims professional exploring automated processing capabilities, or a technology leader implementing AI solutions, this guide offers the insights and frameworks necessary to harness artificial intelligence's transformative potential while addressing the ethical, regulatory, and operational considerations that ensure responsible implementation.

The future of insurance is being written now, and artificial intelligence is the pen. This book ensures you're equipped to help write that future rather than simply react to it.

Chapter 1: Introduction to AI in Insurance

Artificial intelligence represents computer systems that perform tasks traditionally requiring human-like intelligence and learning capabilities. In insurance, AI transforms four key areas. First, it enhances decision-making through data-driven insights and predictive analytics, enabling more accurate risk assessments. Second, it automates routine insurance tasks, significantly increasing operational efficiency while reducing manual errors. Third, AI improves customer experience by providing personalized services and faster response times to inquiries. Finally, it enables continuous learning from new data patterns, allowing insurers to adapt quickly to changing market conditions and emerging risks.

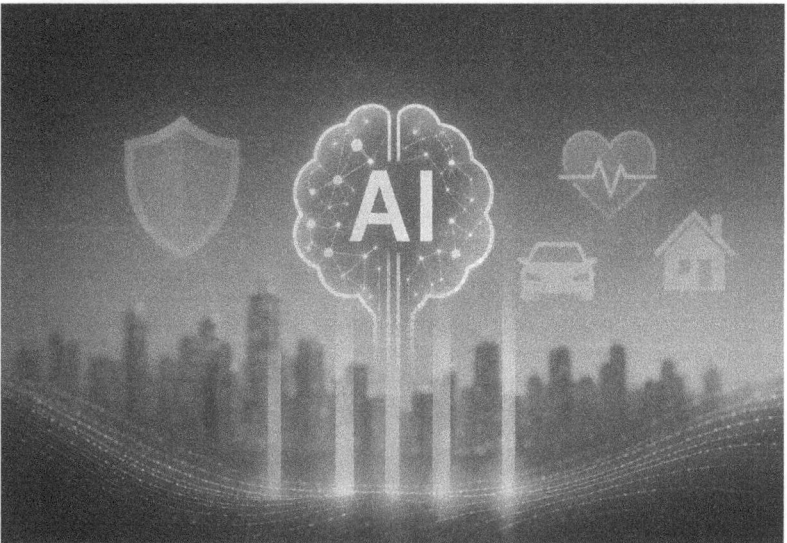

The insurance sector leverages four primary AI technologies. Machine learning enhances underwriting accuracy by analyzing vast datasets to predict risk patterns more effectively than traditional methods. Natural language processing revolutionizes customer service by automatically understanding and responding to customer inquiries, enabling 24/7 support capabilities. Robotic process automation streamlines repetitive tasks such as claims processing, data entry, and policy administration, dramatically increasing efficiency. Computer vision supports fraud detection by analyzing images and documents for anomalies, inconsistencies, or suspicious patterns that human reviewers might miss, protecting insurers from fraudulent claims.

The insurance industry has undergone a dramatic transformation from manual to digital processes. Originally, insurance relied heavily on paper-based workflows that were time-consuming, error-prone, and resource-intensive. The adoption of digital platforms marked a pivotal shift, enhancing accuracy and operational efficiency across all insurance functions. This digital transformation has streamlined workflows, reduced processing errors, and significantly accelerated claim processing times. Modern digital insurance solutions enable real-time data processing, automated decision-making, and seamless customer interactions, setting the foundation for today's AI-driven innovations that further optimize these processes.

Several key milestones have shaped insurance digitization. Online policy management revolutionized customer accessibility, allowing digital policy administration and improving convenience. Telematics in auto insurance enabled real-time monitoring of driving behavior, transforming risk assessment through actual usage data and personalized premium calculations. Predictive analytics empowered insurers to forecast risks and customer behavior patterns, improving underwriting accuracy and decision-making processes. The integration of AI-driven systems into core functions automated complex processes like claims processing and enhanced customer service efficiency, creating the intelligent, responsive insurance ecosystem we see today.

Data and automation have fundamentally reshaped the insurance industry across three dimensions. Faster processing through automation has accelerated workflows, reducing processing time from weeks to minutes and improving overall operational efficiency. Enhanced risk assessment leverages comprehensive data collection to enable more accurate risk evaluation, allowing for tailored and informed decision-making based on real-time insights. Personalized insurance products have emerged through automation that supports customized offerings meeting individual customer needs effectively. This data-driven approach enables insurers to move from one-size-fits-all products to precisely tailored solutions that better serve diverse customer segments.

AI-driven and traditional analytical approaches differ fundamentally in their capabilities and adaptability. AI-driven analytics utilizes advanced algorithms and machine learning techniques to recognize dynamic patterns in data, continuously learning and adapting to new information. These systems can identify complex, non-linear relationships and evolving trends that traditional methods might miss. In contrast, traditional analytics relies on static models and historical data for analysis without adaptive learning capabilities. Traditional approaches use predetermined rules and statistical models that remain constant over time, making them less responsive to changing conditions and emerging patterns in the data landscape.

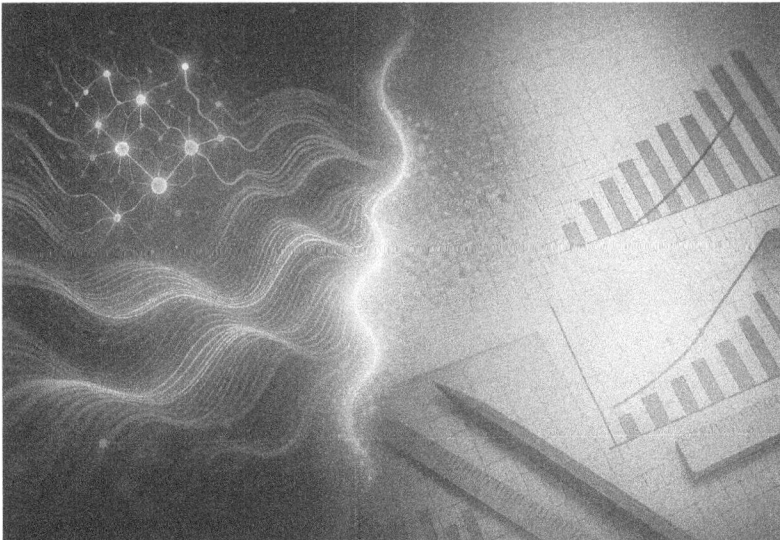

AI delivers significant advantages across core insurance functions. Faster processing accelerates insurance workflows by automating risk assessment, underwriting decisions, and claims handling efficiently, reducing processing times from days to hours. Improved accuracy enhances precision by minimizing human error in evaluating risks and processing claims, leading to more consistent and reliable outcomes. Advanced fraud detection enables AI systems to identify fraudulent activities swiftly by analyzing patterns and anomalies in claims data that human reviewers might overlook. Customer personalization allows AI to enable tailored insurance services by personalizing offers and communication based on individual customer data and preferences.

Both AI and traditional analytics face several significant challenges. Data quality issues remain paramount, as poor data quality undermines the accuracy and reliability of both AI and traditional analytics models, leading to flawed insights and decisions. Model transparency concerns arise particularly with complex AI models that challenge trust and interpretability for decision-makers and regulators. Ethical challenges emerge from biases and fairness issues in analytics that can impact real-world outcomes and customer treatment. Over-reliance risks develop when excessive dependence on algorithms reduces human oversight and critical thinking in analytics, potentially missing important contextual factors that require human judgment.

Leading insurers are implementing AI across multiple operational areas. Claims automation enables faster and more accurate processing of insurance claims through intelligent systems, reducing manual effort and errors while improving customer satisfaction. AI-powered chatbots provide 24/7 customer support, dramatically improving response times and customer satisfaction for insurers while reducing operational costs. Dynamic pricing models leverage AI to analyze vast datasets, creating sophisticated pricing models that optimize insurance premiums based on real-time risk assessments and market factors. Predictive maintenance in asset insurance helps predict maintenance needs for insured assets, reducing risks and costs through proactive interventions before failures occur.

AI drives significant enhancements in products and customer experience across three key areas. Personalized insurance offerings utilize AI to customize insurance products that fit individual customer needs and preferences precisely, moving beyond generic policies to tailored solutions. Rapid claims settlement accelerates the claims process, enabling faster and more accurate settlements for customers through automated assessment and processing systems. 24/7 customer engagement through AI-powered chatbots provides continuous, real-time support, improving customer satisfaction by ensuring help is always available. These improvements create competitive advantages while building stronger customer relationships and loyalty through enhanced service delivery.

AI adoption varies significantly across global regions due to several factors. Regulatory environment impact shows how regulatory policies shape the speed at which AI technologies are adopted across different regions worldwide, with some markets encouraging innovation while others impose stricter controls. Technological infrastructure plays a crucial role, as regions with advanced technological infrastructure experience faster AI adoption and integration capabilities. Market maturity disparities reveal that mature insurance markets lead in AI adoption while emerging markets face challenges in catching up due to resource constraints, regulatory frameworks, and technological readiness. These disparities create varied competitive landscapes globally.

Insurance regulations significantly impact AI implementation across three critical areas. Data privacy regulations enforce strict controls to protect consumers' personal information used in AI insurance applications, requiring robust data governance and security measures. Algorithm transparency rules require clarity in AI algorithms to ensure fairness and accountability in insurance decisions, mandating explainable AI systems. Consumer protection regulations safeguard customers from bias and unfair treatment in AI-driven insurance processes, ensuring that automated decisions don't discriminate against protected groups. These regulatory frameworks create both challenges and opportunities for insurers implementing AI solutions while protecting consumer interests.

Ethical challenges in AI implementation center on three fundamental issues. Transparency in AI systems remains problematic as these systems often lack clear explainability, making transparency crucial for building user trust and accountability among stakeholders. Bias in AI represents a significant concern, as algorithms can introduce biases that result in unfair outcomes and discrimination if not properly identified and managed throughout the development process. Ensuring fairness requires insurers to proactively address ethical challenges to maintain trust in AI-driven decisions and comply with regulatory requirements. These ethical considerations are essential for sustainable AI adoption and maintaining industry reputation.

Insurers are implementing comprehensive strategies to ensure compliance and ethical AI use. Auditing frameworks enable systematic reviews of AI systems for compliance and ethical standards, establishing regular assessment protocols and documentation requirements. Bias detection tools help insurers identify and mitigate unfair biases within AI models before deployment and during ongoing operations. Stakeholder engagement ensures transparency and accountability in ethical AI deployment by involving diverse perspectives from customers, regulators, and internal teams. These proactive measures demonstrate corporate responsibility while building trust with customers and regulators, creating sustainable foundations for AI implementation.

Three emerging AI technologies will significantly impact insurance's future. Explainable AI enhances transparency and trust by making AI decision processes understandable to humans, addressing regulatory requirements and customer concerns. Augmented intelligence supports human decision-making by combining AI capabilities with human expertise, creating powerful hybrid systems that leverage both computational power and human judgment. Edge computing enables faster data processing near the source, improving real-time AI applications in insurance such as instant claims assessment and immediate risk evaluation. These technologies will enable more sophisticated, transparent, and responsive insurance operations.

Insurers should focus on four strategic priorities when investing in AI. Enhancing data capabilities through improved data collection and analysis is crucial for effective AI integration in insurance operations, forming the foundation for all AI applications. Fostering innovation culture by creating environments that encourage experimentation and creativity accelerates AI adoption throughout insurance companies. Regulatory alignment ensures AI initiatives comply with regulatory requirements while protecting customer interests and maintaining operational licenses. Customer-centric AI solutions focus on developing applications that enhance customer experience, strengthening insurer-customer relationships and creating competitive advantages in the marketplace.

The competitive landscape will evolve through four key trends. Increased automation will enable insurers to automate more processes, improving efficiency and reducing operational costs while maintaining service quality. New product development will accelerate as artificial intelligence enables insurers to create innovative insurance products tailored to emerging customer needs and market opportunities. Collaboration with insurtech startups will foster technological advancements and novel services, combining traditional insurance expertise with cutting-edge technology. Intensified competition will result from AI adoption escalating competition among insurers as they leverage technology for market advantage, requiring strategic differentiation.

AI represents a transformative force that will define the insurance industry's future. AI enhances efficiency by streamlining insurance processes, reducing manual work, and speeding up claims handling for superior customer service. Improved accuracy results from AI's ability to analyze large datasets and detect risks more effectively than traditional methods. Customer experience receives significant boosts through AI personalization of customer interactions, offering tailored services and faster responses to inquiries. Insurers adopting AI strategically will gain competitive advantages by driving innovation and staying ahead in the evolving insurance market, making AI adoption not just beneficial but essential for future success.

Chapter 2: Data Infrastructure and Readiness

Data lakes represent a fundamental shift in how insurance companies store and manage information. Unlike traditional databases, data lakes accommodate raw, diverse data in native formats, providing unprecedented flexibility and scalability. For insurance applications, this means integrating multiple data sources seamlessly - from policy management systems to customer communications and external risk databases. The comprehensive data integration capability enables insurers to gain holistic insights across all business functions. Enhanced data access supports sophisticated risk analysis, leading to more accurate underwriting and pricing decisions. Additionally, the rich insights derived from well-structured data lakes significantly improve customer service experiences through personalized offerings and faster claim processing.

Insurance-specific schemas serve as the architectural foundation for organizing complex insurance data effectively. These schemas are purpose-built to accommodate the unique data categories essential to insurance operations: policy information, claims data, customer profiles, and comprehensive risk assessments. The strategic design of these schemas enables seamless querying capabilities, allowing analysts and underwriters to access relevant information quickly and efficiently. Well-designed schemas enhance insurance data analytics capabilities by providing consistent data structures that support advanced analytical tools and machine learning algorithms. This organizational approach transforms raw data into actionable intelligence, supporting better business decisions across underwriting, claims processing, and customer relationship management.

Structured data lakes enable the organized storage of large volumes of diverse insurance data, creating a foundation for advanced analytical capabilities. When properly architected, these systems support sophisticated machine learning applications that can identify insurance trends, detect patterns, and predict outcomes with remarkable accuracy. Machine learning algorithms applied to data lake environments help insurers make better decisions by analyzing historical patterns and real-time information. The predictive analytics capabilities based on comprehensive data lakes significantly improve risk assessment processes, enhance underwriting accuracy, and increase claims processing efficiency. This data-driven approach enables insurers to move from reactive to proactive decision-making, optimizing both operational efficiency and customer outcomes.

Insurance companies generate and collect data from numerous sources, creating both opportunities and challenges for effective data management. Structured insurance data originates from well-organized sources including policy administration systems, claims databases, financial records, and regulatory reporting systems. This data follows predictable formats and can be easily categorized and analyzed. In contrast, unstructured insurance data encompasses diverse formats including customer emails, correspondence, scanned documents, social media interactions, photos from claims adjusters, and sensor-generated files from connected devices. The integration of these varied data sources requires sophisticated processing capabilities but offers comprehensive insights into customer behavior, risk factors, and operational efficiency opportunities.

Effective integration of structured and unstructured data requires sophisticated techniques and technologies. Data parsing serves as the foundation, breaking down complex, unstructured data into structured formats that can be easily integrated and analyzed alongside traditional database information. Natural Language Processing technologies enable machines to understand, interpret, and analyze human language data effectively, extracting meaningful insights from customer communications, claim descriptions, and external sources. Metadata tagging adds descriptive information to all data types, significantly enhancing searchability and usability for analytics applications. These techniques work together to create a unified data environment where diverse information sources contribute to comprehensive analytical insights and improved decision-making processes.

Data harmonization presents several significant challenges that must be addressed systematically. Data format inconsistencies create substantial obstacles when attempting to merge information from different systems and sources effectively. Large volumes of data combined with quality issues impact both the accuracy and efficiency of harmonization processes, potentially leading to flawed analytical results. However, proven solutions exist for these challenges. Implementing standardized schemas across all data sources improves consistency and reliability significantly. Rigorous data cleansing processes ensure information accuracy and completeness. Additionally, utilizing integration platforms that support multiple data formats enables seamless data merging and management, creating a unified environment where diverse data sources work together harmoniously to support business objectives.

Data accuracy and reliability form the cornerstone of effective insurance operations, directly impacting business outcomes and regulatory compliance. Errors in data or outdated information can lead to poor risk assessment, resulting in incorrect pricing, inadequate coverage, or inappropriate claims decisions that ultimately cause significant financial losses. These data quality issues can cascade through the organization, affecting underwriting accuracy, claims processing efficiency, and customer satisfaction levels. Conversely, maintaining high data accuracy standards enhances trust among stakeholders and ensures compliance with regulatory requirements. Accurate, reliable data supports faster decision-making processes, reduces operational risks, and builds customer confidence in the insurance provider's ability to deliver fair, consistent service across all interactions and transactions.

Robust governance frameworks provide the essential structure for maintaining data integrity and implementing reliable management processes throughout the organization. These frameworks establish clear standards and protocols that ensure consistent data handling across all departments and functions. A comprehensive governance approach defines specific roles and responsibilities for data management, ensuring that privacy requirements are met and security protocols are maintained consistently. The governance structure guides uniform data management practices across underwriting and claims departments, eliminating inconsistencies that could compromise analytical accuracy. Effective governance also includes regular auditing processes, quality assurance measures, and continuous improvement protocols that adapt to changing regulatory requirements and business needs while maintaining the highest standards of data integrity.

Strong governance frameworks directly impact the effectiveness of underwriting decisions and claims processing operations. Effective governance ensures consistent adherence to established policies and procedures, significantly enhancing accuracy in both underwriting assessments and claims processing workflows. High-quality, reliable data supports faster and more consistent underwriting decisions while reducing errors and processing delays that can frustrate customers and increase operational costs. Efficient claims handling and underwriting processes, supported by robust data governance, boost customer satisfaction and build trust in the insurance provider's services. Additionally, enhanced processes minimize operational risks by ensuring smoother workflows, reducing the likelihood of compliance issues, and creating predictable, reliable outcomes that support long-term business sustainability and growth.

IoT devices and wearables are revolutionizing insurance by providing unprecedented access to real-time behavioral and risk data. Telematics sensors installed in vehicles collect comprehensive driving data including speed, acceleration, braking patterns, and route information, enabling insurers to assess individual risk profiles accurately and develop personalized premium structures based on actual driving behavior rather than demographic assumptions. Health wearables continuously track vital signs, activity levels, sleep patterns, and other health indicators, providing insurers with detailed insights into customer health behaviors and lifestyle choices. These continuous data streams from connected devices help insurers develop more sophisticated understanding of risk profiles, enabling more accurate pricing, proactive risk management, and improved decision-making across all insurance products and services.

Successfully implementing real-time data ingestion requires robust technological infrastructure and strategic architectural decisions. Streaming data platforms form the backbone of real-time processing, enabling continuous data flow from multiple sources while providing immediate processing capabilities for instant insights and decision-making. Application Programming Interfaces (APIs) facilitate seamless real-time data access and integration from diverse sources, ensuring efficient data capture and transmission between systems and devices. Edge computing represents a critical advancement, processing data near its source to significantly reduce latency and enhance processing speed. This distributed computing approach enables immediate analysis and response to critical data points, supporting real-time risk assessment, dynamic pricing adjustments, and proactive customer engagement strategies that were previously impossible with traditional batch processing systems.

Real-time data capabilities enable transformative applications in risk assessment and customer engagement that fundamentally change how insurance operates. Dynamic pricing powered by real-time data allows businesses to adjust premium rates instantly based on current risk conditions and market dynamics, creating more accurate and competitive pricing strategies. Personalized policies can be crafted using real-time insights, enabling tailored coverage options that match individual customer needs and risk profiles precisely. Proactive risk mitigation becomes possible through real-time alerts and monitoring, allowing insurers to prevent losses by addressing potential issues before they escalate into claims. Enhanced customer communication through real-time data access fosters stronger relationships, enables quicker issue resolution, and supports more responsive customer service that builds loyalty and satisfaction.

In conclusion, robust data infrastructure serves as the essential foundation for advancing insurance operations and supporting increasingly complex analytical requirements. The integration of diverse data sources, combined with strong governance frameworks, ensures that insurers can rely on accurate, consistent insights for critical business decisions. Quality governance practices guarantee data reliability and regulatory compliance while supporting operational efficiency. Real-time IoT data usage represents the future of insurance analytics, enabling superior risk management capabilities and more engaging customer experiences. By implementing these comprehensive data management strategies, insurance companies can transform their operations, improve customer satisfaction, reduce risks, and maintain competitive advantages in an increasingly data-driven marketplace. Success requires commitment to both technological advancement and governance excellence.

Chapter 3: AI Techniques and Tools in Insurance

Machine learning represents a fundamental shift in how insurance companies approach risk assessment. Unlike traditional static models that rely on predetermined variables, ML algorithms continuously learn from new data, identifying patterns that human analysts might miss. The technology excels at processing massive insurance datasets, uncovering hidden correlations between seemingly unrelated factors that influence risk. This enhanced data analysis capability translates directly into improved underwriting accuracy, enabling insurers to make more precise pricing decisions and reduce adverse selection. Most importantly, machine learning enables dynamic risk evaluation that adapts to changing conditions in real-time, moving beyond the limitations of traditional static risk assessment methods that may become outdated quickly.

Insurance companies employ three primary categories of machine learning algorithms, each serving distinct purposes in risk modeling. Regression algorithms are particularly valuable for predicting numeric outcomes, such as claim amounts or policy values, by analyzing relationships between multiple variables in historical data. Classification algorithms excel at categorizing risks into predefined groups, helping insurers determine appropriate risk tiers and policy terms based on customer characteristics and behavior patterns. Clustering algorithms offer perhaps the most innovative approach, discovering hidden groupings within insurance data without predetermined labels. This unsupervised learning technique can reveal new customer segments or risk patterns that weren't previously recognized, providing insurers with fresh insights for product development and risk management strategies.

Implementing machine learning in risk modeling brings significant advantages but also presents notable challenges that insurers must carefully navigate. The enhanced predictive accuracy achieved through ML can dramatically improve portfolio performance and reduce unexpected losses. Automation of complex analytical tasks saves considerable time and reduces manual errors that can occur in traditional risk assessment processes. However, data quality remains a critical challenge – machine learning models are only as reliable as the data they're trained on, making data governance essential. Additionally, model interpretability poses regulatory and business challenges, as insurance companies must be able to explain their decisions to regulators and customers while ensuring compliance with fair lending and anti-discrimination laws.

Natural language processing is transforming document management in insurance by automating traditionally labor-intensive processes. NLP technology can extract essential information from policies, claims documents, and customer communications without manual intervention, dramatically improving both accuracy and processing speed. This automation significantly reduces the human workload required for document review, minimizing the potential for human error while freeing up staff for higher-value activities. The technology excels at identifying key data points, dates, amounts, and other critical information from unstructured text. Perhaps most importantly, NLP accelerates processing times substantially, enabling faster decision-making and improved workflow efficiency. This speed improvement directly translates to better customer service and reduced operational costs for insurance companies.

AI-powered chatbots and virtual assistants are revolutionizing customer support in the insurance industry by providing continuous, high-quality service capabilities. These systems operate 24/7 without downtime, ensuring customers can access support whenever they need it, regardless of time zones or business hours. Chatbots efficiently handle routine customer queries, helping users quickly find relevant policy information, coverage details, and answers to common questions. Virtual assistants excel at guiding customers through complex policy options, breaking down complicated insurance terminology into understandable language. This technology significantly enhances the overall customer experience by providing prompt, personalized responses. The combination of immediate availability, consistent service quality, and personalized interaction capabilities leads to improved customer satisfaction and reduced support costs for insurance companies.

Natural language processing plays a crucial role in enhancing fraud detection and ensuring regulatory compliance within insurance operations. NLP systems analyze textual data from claims, communications, and documents to identify suspicious patterns and potential fraud indicators that might escape human detection. The technology can spot inconsistencies in language patterns, unusual terminology, or narrative elements that suggest fraudulent activity. For regulatory compliance, NLP helps monitor all textual content to ensure adherence to industry policies and regulatory standards, automatically flagging potential violations or compliance issues. By integrating NLP into risk management processes, insurance companies can proactively detect fraud before claims are paid and ensure consistent policy adherence across all operations, ultimately protecting both the company and legitimate customers from fraudulent activities.

Computer vision technology is revolutionizing insurance claims processing by enabling automated analysis of customer-submitted photographs and documents. Image recognition systems can automatically assess damage from photos submitted by policyholders, identifying the type, extent, and estimated cost of repairs needed. This technology improves accuracy by providing consistent, objective damage assessment that reduces human bias and error. The automated analysis significantly accelerates claim validation processes, reducing the time between claim submission and approval. Customers benefit from faster claim resolutions, while insurers reduce processing costs and improve operational efficiency. This technology is particularly valuable for property damage, auto insurance claims, and natural disaster assessments, where visual evidence is critical for determining coverage and compensation amounts.

Computer vision systems provide powerful capabilities for detecting document forgeries and identifying anomalies that indicate potential fraud. These tools analyze documents pixel by pixel, identifying inconsistencies in fonts, formatting, image quality, and other subtle indicators that suggest alterations or forgeries. The technology can detect sophisticated manipulation techniques that might fool human reviewers, including digital alterations, copied signatures, or inserted text. By identifying these anomalies early in the claims process, insurance companies can prevent fraudulent claims from being processed and maintain document authenticity standards. This proactive approach to fraud prevention protects both the insurance company's financial interests and helps keep premium costs reasonable for honest policyholders by reducing the overall cost of fraud in the system.

Automated document classification is streamlining underwriting processes by efficiently sorting and categorizing application materials without human intervention. This technology can automatically identify different document types – medical records, financial statements, property appraisals, driving records – and route them to appropriate review processes. The automation significantly speeds up underwriting by eliminating manual sorting and classification tasks that traditionally consume considerable time. By automating document handling, risk assessment becomes both quicker and more accurate, as documents are processed consistently and nothing falls through administrative cracks. This streamlined workflow leads to faster policy approvals and quicker policy issuance to clients, improving customer satisfaction while reducing operational costs. The technology also ensures that all required documentation is present before underwriting begins.

Chapter 3: AI Techniques and Tools in Insurance

Predictive modeling enables insurance companies to forecast both claim frequency and severity with unprecedented accuracy using historical data and advanced statistical techniques. These models analyze patterns in past claims to predict how many claims are likely to occur in future periods, helping insurers prepare operationally and financially. Claim severity estimation focuses on forecasting the likely financial cost of individual claims, enabling better budgeting and reserve setting. This dual approach to prediction – both frequency and severity – provides insurers with comprehensive insights for risk management decisions. Accurate claim forecasts assist organizations in making informed choices about pricing, coverage limits, reinsurance purchases, and capital allocation. The predictive insights also help insurers identify emerging risks and adjust their strategies proactively rather than reactively.

Generative AI is enabling unprecedented personalization in insurance offerings by creating tailored products that match individual customer needs and risk profiles. This technology analyzes comprehensive customer data – including demographics, behavior patterns, risk factors, and preferences – to design customized insurance policies rather than forcing customers into rigid, one-size-fits-all products. AI-powered customer risk profiling goes beyond traditional underwriting by considering a broader range of factors and their interactions, creating more nuanced and accurate risk assessments. This personalization benefits both insurers and customers: insurers can price more accurately and reduce adverse selection, while customers receive coverage that better matches their actual needs and risk levels.

The result is improved customer satisfaction, better risk management, and more competitive positioning in the marketplace.

Predictive insights are transforming customer engagement by enabling insurers to provide proactive, personalized service that anticipates customer needs. Data-driven customer engagement uses predictive analytics to tailor offers, recommendations, and communications for each individual customer, resulting in more relevant interactions and higher satisfaction levels. Insurers can now provide proactive risk advice, warning customers about potential hazards before they result in claims, such as weather alerts, maintenance reminders, or safety recommendations based on individual risk profiles. This proactive approach helps customers avoid losses while reducing claim costs for insurers. The enhanced personalization and proactive service leads to improved customer retention rates, as customers appreciate receiving valuable, personalized support rather than generic communications. This approach transforms insurance from a reactive service to a proactive partnership.

In conclusion, artificial intelligence is fundamentally transforming the insurance industry across multiple dimensions. Machine learning, natural language processing, and computer vision represent the core technologies driving this transformation, each contributing unique capabilities to traditional insurance operations. These AI tools deliver substantial efficiency and accuracy improvements, enabling insurers to process information faster and make better decisions with reduced human error. The enhanced customer experience created by AI tools – through personalized services, faster claim resolutions, and 24/7 support – is becoming a key competitive differentiator in the insurance

28

marketplace. Finally, predictive analytics provides superior risk management capabilities, helping insurers identify, assess, and mitigate risks more effectively in underwriting, fraud detection, and portfolio management. The integration of these technologies positions forward-thinking insurers for sustained competitive advantage.

Chapter 4: AI-Driven Underwriting

Our first major topic focuses on Automated Underwriting Workflows. This section will demonstrate how AI is fundamentally changing the way insurance companies process applications, assess risks, and make underwriting decisions. We'll explore the mechanisms behind automation and the significant benefits it brings to both insurers and customers through enhanced efficiency and consistency.

AI streamlines the underwriting process through three key mechanisms. First, automation of routine tasks involves AI handling repetitive underwriting activities like data extraction, which minimizes manual labor and significantly reduces human errors. Second, risk evaluation enhancement allows AI to improve accuracy in assessing risk by analyzing large, complex datasets quickly and effectively, identifying patterns that human underwriters might miss. Third, the accelerated underwriting cycle means automation speeds up the entire process, reducing approval times and improving overall operational efficiency.

The integration of machine learning models represents a crucial advancement in underwriting decision-making. These sophisticated models excel at complex data analysis, processing and analyzing large, complex datasets efficiently to extract meaningful insights that inform better decisions. Enhanced risk prediction capabilities allow these models to predict risk with higher accuracy by identifying subtle patterns that traditional statistical methods often overlook. Most importantly, machine learning provides robust support for underwriting decisions by delivering data-driven insights that significantly improve the quality and consistency of risk assessments.

The benefits of automated underwriting workflows center on four key improvements. Enhanced processing speed through automation significantly reduces workflow turnaround times and increases overall productivity. Minimized human error is achieved as automated systems eliminate manual input mistakes and ensure greater data accuracy throughout the process. Consistent underwriting application ensures that guidelines are applied uniformly across all cases, enhancing reliability and regulatory compliance. Finally, improved customer experience results from faster and more accurate processing, leading to better service delivery and increased customer satisfaction levels.

Dynamic Pricing Models with AI represent our second major topic. This section explores how artificial intelligence enables insurers to move beyond static pricing structures to real-time, data-driven pricing strategies. We'll examine how AI analyzes continuous data streams to adjust pricing dynamically based on current risk assessments and individual customer profiles.

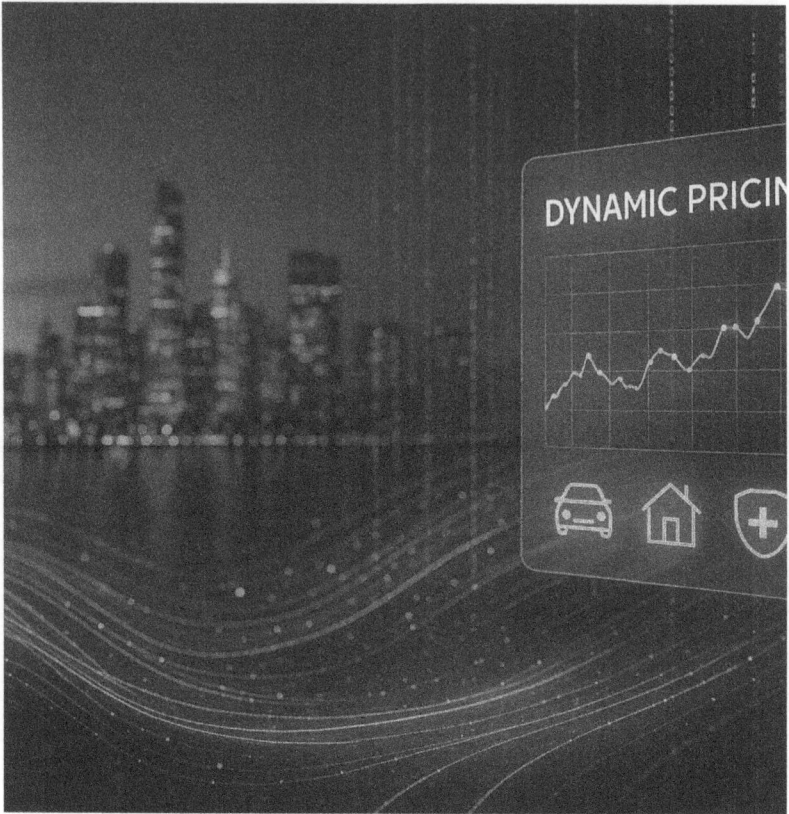

Real-time risk assessment using AI operates through three interconnected processes. AI monitoring of data streams involves continuous analysis of incoming data to identify emerging risks in real time, providing unprecedented visibility into changing conditions. Instant risk updates ensure that risk assessments are refreshed immediately based on real-time data analysis, eliminating delays in recognizing new threats or opportunities. Efficient pricing adjustments allow insurers to use these AI insights to modify pricing and coverage terms efficiently and timely, ensuring that rates accurately reflect current risk levels.

Personalized pricing strategies represent a significant advancement in insurance customization. AI-driven customer analysis evaluates individual customer behavior and risk profiles to enable precise pricing adjustments tailored specifically to each customer's unique circumstances. This granular approach allows for customized pricing plans that optimize value proposition by reflecting the distinct needs and risk factors of each individual customer. The result is more accurate risk-based pricing that benefits both low-risk customers through reduced premiums and insurers through better risk selection.

Continuous model updates based on new data ensure that AI pricing systems remain current and effective. AI model learning involves pricing models that continuously absorb fresh data to enhance prediction accuracy over time, creating a self-improving system. Adaptation to risk changes allows models to dynamically adjust to evolving risk environments, ensuring that pricing remains relevant and reliable as market conditions, customer behaviors, and external factors change. This continuous learning approach maintains the effectiveness of pricing strategies in dynamic market conditions.

Behavioral Data From Telematics and Wearables forms our third major topic. This section examines how modern sensor technologies provide unprecedented insights into customer behavior, enabling more accurate risk assessment and personalized insurance products. We'll explore how this data transforms traditional underwriting approaches through real-world behavioral evidence.

Collecting behavioral data via telematics devices revolutionizes auto insurance underwriting. Telematics devices functionality involves recording detailed driving behavior and vehicle usage data in real time, providing comprehensive monitoring of actual driving patterns rather than relying on demographic assumptions. This improved risk assessment capability allows collected data to enable more accurate risk evaluation in auto insurance underwriting, significantly reducing uncertainty and enabling more precise premium calculations based on actual behavior rather than statistical averages.

Wearable technology provides comprehensive health and lifestyle insights for life and health insurance. Continuous health monitoring through wearable devices tracks vital health metrics such as heart rate, activity levels, and sleep patterns in real time, providing ongoing health status updates. Lifestyle behavior tracking records comprehensive behavioral patterns including physical activity levels, sleep quality, and daily routines to provide detailed lifestyle insights. Insurance risk evaluation utilizes this wearable data to more accurately assess life and health insurance risks and create personalized policies that reflect actual health status and lifestyle choices.

The impact of behavioral data on underwriting accuracy is transformative across three key areas. Accurate risk profiling through behavioral data integration enhances precision in identifying individual risk levels, leading to better-informed underwriting decisions based on actual behavior rather than assumptions. Improved pricing models utilize behavioral insights to allow insurers to tailor pricing that reflects real-time risk and actual customer behavior patterns. Enhanced customer engagement through rewarding safer behavior and healthy lifestyles encourages positive customer interaction, builds loyalty, and creates incentives for risk reduction.

Bias Detection in Underwriting Algorithms represents our final major topic. This critical section addresses fairness and compliance challenges in AI-driven underwriting systems. We'll examine common sources of algorithmic bias and explore techniques for identifying, mitigating, and preventing unfair discrimination in automated underwriting processes.

Common sources of bias in AI models stem from three primary areas that underwriting teams must actively address. Training data imbalances often originate from unbalanced datasets that misrepresent certain demographic groups or geographic regions, leading to skewed risk assessments. Flawed assumptions in model design can introduce systematic bias when incorrect or oversimplified assumptions about risk factors are built into the algorithmic framework. Historical prejudices embedded in datasets can perpetuate past unfairness in AI underwriting models, requiring careful data curation and bias testing.

Techniques for bias identification and mitigation involve systematic approaches to ensure fairness. Fairness audits systematically evaluate AI systems to uncover and address bias in underwriting processes through regular testing across demographic groups and risk categories. Algorithmic transparency involves increasing visibility into decision-making processes, helping stakeholders understand how decisions are made and detect potentially biased behaviors. Adversarial testing challenges AI models with difficult and edge-case inputs specifically designed to identify bias and improve model robustness against unfair discrimination.

Ensuring fairness and regulatory compliance requires ongoing organizational commitment. Governance frameworks establish structured approaches that guide AI development to consistently meet legal and ethical standards, providing clear guidelines for bias prevention and detection. Continuous monitoring ensures that AI models maintain compliance and fairness for all customer groups through regular testing, validation, and adjustment processes. This ongoing oversight is essential for maintaining public trust and regulatory compliance in AI-driven underwriting systems.

The future of AI-driven underwriting promises significant improvements across four key dimensions. Efficiency improvements through AI streamline the underwriting process, reducing processing time and operational costs significantly while maintaining quality standards. Enhanced accuracy allows AI models to analyze data precisely, minimizing errors and improving risk assessments beyond human capabilities. Personalized underwriting enables tailored insurance solutions by factoring individual customer data and behaviors into risk assessment and pricing decisions. Finally, fairness and transparency support more equitable underwriting decisions by

Chapter 4: AI-Driven Underwriting

reducing bias and increasing transparency in decision-making processes, ensuring ethical AI implementation in insurance.

Chapter 5: Risk Profiling and Scoring

Risk profiling combines two complementary methodologies to create comprehensive assessments. Qualitative assessments leverage expert judgment and descriptive analysis, providing valuable context and nuanced understanding of complex risk factors that numbers alone cannot capture. Quantitative models employ statistical techniques and numerical data to deliver precise, measurable risk evaluations that enable consistent comparison across different scenarios. The most effective risk management strategies integrate both approaches, creating a balanced framework that combines the precision of data-driven analysis with the contextual understanding that human expertise provides. This dual approach ensures more robust and reliable risk evaluations.

Risk scoring serves as the critical bridge between raw risk data and actionable management decisions. The scoring process converts both qualitative insights and quantitative measurements into standardized, comparable metrics that stakeholders can easily understand and act upon. These scores provide clear prioritization frameworks, helping organizations focus resources on the most significant threats while maintaining awareness of lower-priority risks. Effective scoring systems enhance decision-making by presenting complex risk information in accessible formats, enabling faster response times and more strategic resource allocation. The ultimate goal is transforming data complexity into decision clarity through well-designed scoring methodologies.

Technology integration fundamentally transforms traditional risk assessment practices through three key improvements. Enhanced data processing capabilities allow AI systems to analyze vast volumes of information at unprecedented speeds, identifying patterns and correlations that manual analysis might miss. Improved modeling accuracy emerges from advanced algorithms that can handle complex, multi-variable scenarios with greater precision than traditional statistical methods. Predictive analytics represents the most significant advancement, enabling organizations to anticipate and prepare for potential risks before they materialize. This proactive approach shifts risk management from reactive damage control to preventive strategic planning, ultimately reducing both likelihood and impact of adverse events.

Real-time risk scoring depends on three interconnected technological components working in harmony. IoT sensors continuously collect critical data from various sources, providing the foundation for accurate, up-to-date risk assessments across multiple environments and scenarios. Streaming data platforms ensure seamless data flow, enabling immediate processing and analysis without delays that could compromise response effectiveness. Rapid processing algorithms represent the analytical engine, instantly converting incoming data into actionable risk scores that decision-makers can use immediately. Together, these components create an integrated system capable of providing instant risk intelligence, dramatically reducing the time between risk emergence and management response.

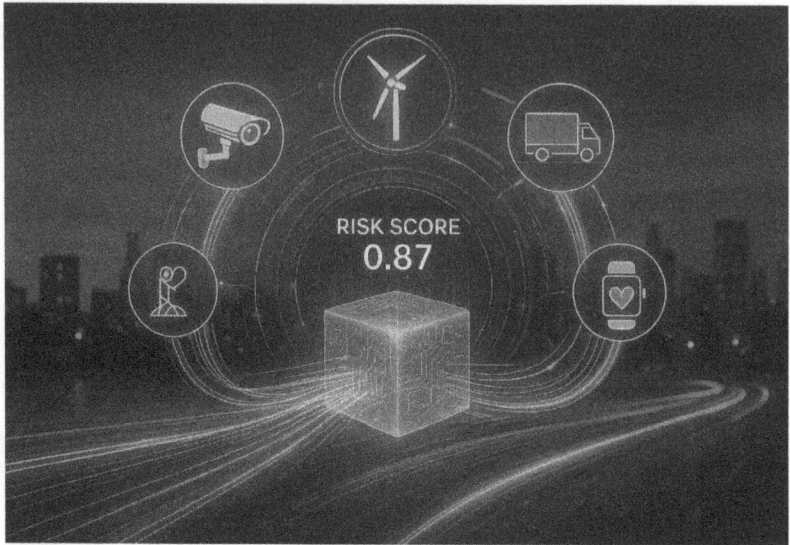

Real-time scoring delivers significant benefits while presenting notable implementation challenges. The primary advantage lies in timely insights and enhanced agility, enabling organizations to respond to emerging risks within minutes rather than hours or days. However, data quality challenges pose ongoing concerns, as real-time systems require consistently high-quality inputs to maintain accuracy in their outputs. Additionally, latency and infrastructure requirements present significant technical hurdles, demanding robust, scalable systems capable of handling continuous data streams without performance degradation. Successfully implementing real-time scoring requires careful balance between speed and accuracy, ensuring that rapid responses remain well-informed and strategically sound.

Real-time risk scoring applications demonstrate practical value across diverse industries and use cases. Financial fraud detection systems can identify suspicious transactions within milliseconds, preventing losses before they occur through immediate account freezes or transaction blocks. Supply chain risk monitoring enables rapid responses to disruptions, minimizing operational impact through alternative sourcing or logistics adjustments. Health risk assessments support critical medical decisions by providing instant patient risk evaluations, enabling healthcare providers to prioritize interventions and optimize treatment protocols. Each application showcases how real-time scoring transforms risk management from reactive to proactive, creating competitive advantages through enhanced responsiveness and improved outcomes.

Machine learning techniques specifically designed for catastrophic event prediction address the unique challenges of rare, high-impact scenarios. Neural networks excel at modeling complex data patterns, learning from historical events and environmental factors to identify subtle precursors to catastrophic occurrences. Ensemble models combine multiple algorithmic approaches, improving prediction accuracy by leveraging the strengths of different methodologies while compensating for individual model weaknesses. Anomaly detection techniques focus on identifying unusual patterns that deviate from normal baselines, providing early warning signals for potential catastrophic events. These approaches work together to enhance predictive capabilities for events that traditional statistical models struggle to anticipate effectively.

AI-driven scenario analysis revolutionizes disaster preparedness through sophisticated simulation and forecasting capabilities. Disaster scenario simulation enables organizations to model diverse catastrophic events, testing response strategies and identifying

44

vulnerabilities before actual crises occur. This capability allows for comprehensive preparedness planning across multiple potential disaster types and severity levels. Risk forecasting leverages AI's ability to analyze complex variables and massive datasets, providing enhanced predictive capabilities that extend far beyond traditional forecasting methods. By processing historical patterns, current conditions, and emerging trends simultaneously, AI systems can identify potential risks with greater accuracy and longer lead times than conventional approaches.

AI enhances risk mitigation strategies through three interconnected capabilities that transform organizational resilience. AI-driven decision making processes complex data to provide tailored mitigation strategies, ensuring responses are precisely calibrated to specific risk profiles and organizational contexts. Adaptive mitigation strategies represent a dynamic approach that evolves with changing risk environments, automatically adjusting protection measures as threats shift or intensify. Improved resilience emerges from AI's ability to anticipate threats and optimize response protocols, creating organizations that are not merely reactive but proactively prepared. This comprehensive approach builds robust defense systems that can withstand and quickly recover from catastrophic events while maintaining operational continuity.

External data integration significantly enhances risk scoring accuracy through diverse information sources that traditional models often overlook. Social media provides real-time public sentiment analysis, enabling instant access to collective opinions and reactions as events unfold globally, offering valuable insights into emerging social and political risks. Behavioral data helps identify new risk patterns early in their development, allowing organizations to develop proactive mitigation strategies before risks fully materialize. The integration of social and behavioral data creates enhanced risk datasets that enrich traditional financial and operational models with dynamic, real-world information. This comprehensive data approach provides more nuanced, contextually aware risk assessments that better reflect actual conditions.

Climate and environmental datasets provide crucial inputs for comprehensive risk assessment strategies in our changing world. These datasets enable organizations to evaluate risks from extreme weather events and natural disasters more effectively, incorporating long-term climate trends and short-term weather patterns into risk models. Environmental monitoring capabilities track long-term ecological changes, supporting sustainable planning initiatives and climate adaptation strategies. By integrating climate science with risk management, organizations can better prepare for environmental challenges while identifying opportunities in the evolving regulatory and market landscape. This data-driven approach to environmental risk assessment ensures that climate considerations become integral to strategic planning and operational decision-making processes.

Geolocation technologies enable precise, context-specific risk analysis that significantly enhances traditional assessment methods. Geolocation data utilization helps identify exact locations of potential risks, providing granular detail that enables more targeted and effective risk management strategies. This precision allows for resource optimization and more accurate threat assessment across different geographic areas. Contextual risk analysis emerges from mapping risks spatially, providing deeper understanding of how location-specific factors influence risk profiles and response strategies. By incorporating geographic context, organizations can develop more nuanced risk management approaches that account for local conditions, regulatory environments, and regional characteristics that significantly impact risk exposure and mitigation effectiveness.

Tailoring AI risk models for niche markets requires specialized approaches that address unique industry characteristics and challenges. Industry-specific customization involves adapting AI algorithms to recognize and weight factors that are particularly relevant to specialized sectors, improving accuracy in identifying industry-specific risks and opportunities. Enhanced risk profiling emerges from focusing on specialized risk pools, creating more precise assessments that reflect the unique dynamics of niche markets. Increased model relevance ensures that AI systems remain effective in specialized contexts with distinct challenges, regulatory requirements, and operational characteristics. This customization process requires deep industry knowledge combined with technical expertise to create truly effective specialized risk assessment tools.

Limited data scenarios present unique challenges that AI can address through innovative approaches designed for data-scarce environments. Transfer learning enables AI models to apply knowledge gained from large, well-documented datasets to improve performance in situations where historical data is limited or unavailable. Synthetic data generation provides a powerful tool for augmenting scarce datasets, creating realistic training data that improves AI model performance in specialized risk pools. These advanced AI methods effectively tackle the fundamental challenge of building robust risk models when traditional data sources are insufficient. The combination of transfer learning and synthetic data generation creates viable pathways for developing sophisticated risk assessment capabilities even in previously underserved market segments.

Case studies demonstrate the practical benefits and real-world applications of AI in specialized risk pool management across various industries. AI-driven risk models consistently show enhanced accuracy in predicting risks for specialized markets, leading to improved overall management efficiency and better resource allocation decisions. Improved decision-making emerges from AI insights that provide deeper understanding of small market environments, enabling organizations to optimize resource allocation and achieve better outcomes with limited resources. Real-world applications illustrate how theoretical AI capabilities translate into practical benefits, providing concrete evidence of value creation in niche markets. These case studies serve as blueprints for organizations considering similar implementations in their specialized risk management contexts.

Advanced AI fundamentally transforms risk profiling by enabling faster, more accurate assessments across diverse applications and industries. The integration of external data sources represents a paradigm shift, improving context-awareness and precision in risk scoring systems through comprehensive information gathering from multiple sources. These technological advances collectively provide more timely and context-aware solutions across diverse risk management applications, creating competitive advantages for organizations that successfully implement these capabilities. The future of risk management lies in the intelligent integration of AI technologies with traditional risk assessment methods, creating hybrid approaches that leverage the best of both human expertise and artificial intelligence capabilities for superior risk management outcomes.

Chapter 6: Fraud Detection and Prevention

We begin with a comprehensive overview of fraud detection and prevention fundamentals. This section establishes the critical foundation for understanding modern fraud challenges and the necessity for proactive management approaches in today's complex threat landscape.

Fraud in insurance and financial sectors encompasses a wide variety of deceptive practices. False claims represent a significant portion, where individuals or organizations submit fabricated or exaggerated damage reports. Identity theft continues to grow, with criminals exploiting personal data to create fraudulent accounts or policies. Financial statement manipulation involves altering documents to obtain unauthorized benefits or coverage. The financial impact is staggering, with billions in losses annually affecting not just businesses but also legitimate customers through increased premiums and fees. Understanding this complexity is essential for developing comprehensive detection strategies that address multiple fraud vectors simultaneously.

Proactive fraud management represents a paradigm shift from reactive to preventive approaches. Early detection systems monitor transactions and behaviors in real-time, identifying suspicious activities before significant damage occurs. This approach enables organizations to implement immediate protective measures, such as account freezes or additional verification requirements. Risk mitigation becomes more effective when potential threats are identified during their initial stages, allowing for targeted interventions. The key advantage is staying ahead of evolving fraud tactics rather than responding after losses have occurred. Organizations implementing proactive measures typically see substantial reductions in fraud-related losses and improved customer trust through enhanced security measures.

The fraud landscape continuously evolves as criminals adapt to new technologies and security measures. Traditional fraud methods are being enhanced with sophisticated digital tools, making detection increasingly challenging. Identity theft has become more complex with the availability of personal data through data breaches and social engineering. Cybercriminals now use advanced techniques including deepfakes, AI-generated documents, and coordinated attacks across multiple channels. The rise of synthetic identities, where fraudsters combine real and fabricated information, presents new challenges for traditional verification methods. Organizations must maintain current threat intelligence and adapt their detection systems accordingly to address these emerging risks effectively.

Anomaly detection in claims and policies represents a sophisticated approach to identifying suspicious patterns that deviate from normal behavior. This section explores statistical and machine learning techniques for recognizing fraudulent activities.

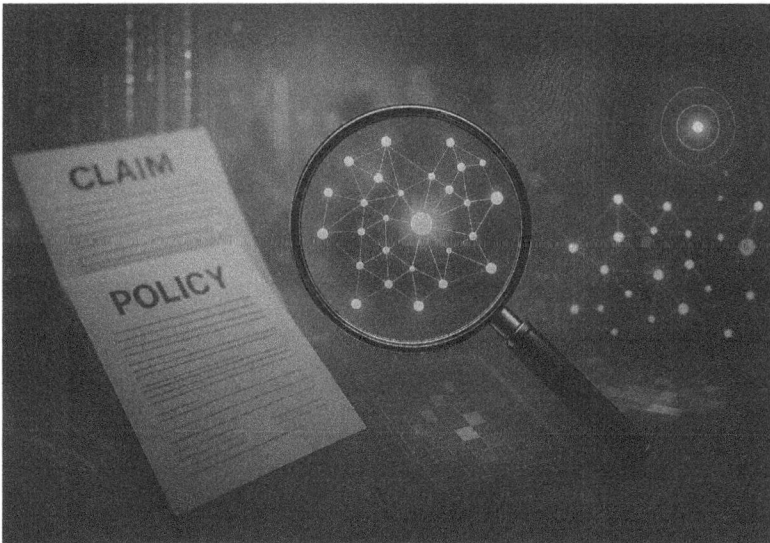

Statistical approaches form the foundation of anomaly detection systems. Data distribution analysis examines the normal patterns within datasets, establishing baselines for typical behavior. When new data points significantly deviate from these established patterns, they are flagged as potential anomalies. Clustering techniques group similar data points together, making outliers more visible and easier to identify. Points that don't fit into any established cluster become candidates for further investigation. Regression analysis helps identify data points that fall outside expected trends or relationships. These statistical methods provide the mathematical foundation for more advanced machine learning approaches, offering interpretable results that fraud investigators can understand and act upon effectively.

Machine learning models enhance claims analysis through sophisticated pattern recognition capabilities. Decision trees create clear, interpretable rules for fraud detection by splitting data based on specific criteria. These trees provide transparent decision-making processes that investigators can follow and understand. Random forests combine multiple decision trees to improve accuracy and reduce the risk of overfitting to specific patterns. This ensemble approach provides more robust predictions by considering multiple perspectives on the same data. Neural networks excel at learning complex, non-linear patterns that traditional statistical methods might miss. They can identify subtle relationships between variables that indicate fraudulent behavior, adapting their detection capabilities as they process more data and learn from new fraud patterns.

Real-world applications demonstrate the effectiveness of anomaly detection systems. Reduced false positives represent a significant operational improvement, as investigators can focus their attention on genuinely suspicious activities rather than investigating legitimate transactions flagged by overly sensitive systems. Insurance companies and financial institutions report substantial improvements in fraud detection rates when implementing these systems. The technology helps identify previously undetected fraud patterns, particularly sophisticated schemes that might evade traditional rule-based systems. Operational efficiency gains result from automated initial screening, allowing human investigators to concentrate on complex cases requiring judgment and expertise. Cost savings emerge through both reduced fraud losses and more efficient use of investigative resources.

Deep learning applications in image and document forgery detection represent cutting-edge technology for verifying the authenticity of submitted materials in insurance claims and financial applications.

Convolutional Neural Networks excel at image authentication through detailed feature analysis. These systems examine pixel-level patterns, lighting consistency, and compression artifacts that may indicate manipulation. CNNs can detect sophisticated forgeries including photoshopped images, digitally altered documents, and deepfake photographs. Automated verification systems process thousands of images quickly, identifying potential forgeries without manual inspection. This technology is particularly valuable for insurance claims involving accident photos, property damage images, and identity verification documents. The systems learn to recognize subtle inconsistencies that human reviewers might miss, including metadata analysis and technical signatures that indicate digital manipulation. Implementation significantly reduces the time required for document verification while improving accuracy rates.

Chapter 6: Fraud Detection and Prevention

Natural Language Processing transforms document verification by analyzing textual content for inconsistencies and potential forgeries. Text extraction techniques automatically pull relevant information from documents, enabling systematic analysis of claims forms, financial statements, and supporting documentation. Advanced algorithms compare extracted information against known patterns and databases to identify discrepancies. Inconsistency detection algorithms analyze writing styles, terminology usage, and document structure to identify potentially fabricated documents. NLP systems can detect when multiple documents allegedly from the same source show different writing patterns or when technical language is used incorrectly. This technology significantly improves claims handling efficiency by automating initial document screening and flagging suspicious materials for human review.

Integration into claims processing workflows streamlines verification processes through automation. Deep learning tools are embedded directly into existing systems, automatically analyzing submitted documents and images as part of the standard claims process. This integration reduces manual effort by eliminating the need for investigators to manually review every submitted document. Staff can focus their expertise on complex cases that require human judgment and investigation skills. The technology accelerates fraud detection by providing immediate analysis results, enabling faster decision-making on legitimate claims while flagging suspicious submissions for detailed review. Organizations report significant improvements in processing speed and accuracy, leading to better customer satisfaction for legitimate claimants and more effective fraud prevention.

Chapter 6: Fraud Detection and Prevention

Cross-channel behavior analysis integrates data from multiple communication channels to provide comprehensive fraud detection capabilities that identify coordinated and organized fraudulent activities.

Multi-source data aggregation combines information from digital platforms, phone interactions, and in-person engagements to create comprehensive user profiles. Digital platform data includes online behaviors, transaction patterns, device information, and application usage statistics. Phone interaction data encompasses call patterns, voice analysis, and communication timing that can reveal suspicious coordination between individuals. In-person engagement data adds behavioral observations and document verification results from face-to-face interactions. This comprehensive approach enables organizations to build detailed risk profiles that consider the full spectrum of customer interactions. Enhanced fraud detection capabilities emerge from correlating patterns across these diverse data sources, revealing sophisticated fraud schemes that might be invisible when examining single channels in isolation.

Cross-channel pattern detection reveals sophisticated fraud schemes operating across multiple communication channels. Fraudsters often use different channels to avoid detection, but analyzing patterns across digital, phone, and in-person interactions can expose these coordinated efforts. The technology identifies suspicious correlations such as multiple individuals using similar scripts across different channels, coordinated timing of claims submissions, or inconsistent information provided through different communication methods. Enhanced fraud detection capabilities result from combining insights across all customer touchpoints. This comprehensive approach is particularly effective against organized fraud rings that deliberately spread their activities across multiple channels to avoid detection. Organizations can identify previously undetectable patterns and respond more effectively to sophisticated threats.

Sophisticated fraud rings operate across multiple channels, making detection complex and requiring advanced analytical approaches. These organized groups often employ coordinated strategies, using different individuals to interact through various channels while maintaining consistent fraudulent objectives. Detection techniques must be equally sophisticated, employing network analysis to identify connections between seemingly unrelated individuals and activities. Advanced pattern recognition algorithms analyze communication patterns, timing correlations, and behavioral similarities across different channels. Protecting organizations from these large-scale fraud operations requires implementing robust detection systems that can identify and disrupt coordinated activities before significant losses occur. Success depends on combining technology with human expertise to investigate and respond to complex fraud networks.

Adaptive fraud scoring engines represent the next generation of fraud detection technology, providing dynamic, real-time risk assessment capabilities that continuously evolve with changing threat patterns.

Dynamic scoring models provide real-time risk assessment capabilities that adapt continuously as new information becomes available. These systems update risk scores instantaneously as additional data arrives, ensuring that fraud detection remains current and effective. Immediate threat response capabilities enable organizations to react quickly to emerging fraud patterns, potentially stopping fraudulent activities in progress. The technology monitors multiple risk factors simultaneously, adjusting scores based on changing behaviors, new transaction patterns, and emerging threat intelligence. Fraud loss reduction results from proactive identification and mitigation of high-risk activities before significant damage occurs. These systems learn from each interaction, continuously improving their accuracy and effectiveness in identifying genuine threats while reducing false positive rates.

Personalization and risk profiling enhance fraud detection accuracy through individual behavior analysis. These systems create detailed profiles for each user, learning normal behavior patterns and identifying deviations that might indicate fraudulent activity. Individual risk profiles consider factors such as transaction history, communication patterns, geographic locations, and timing of activities. This personalized approach significantly improves fraud detection effectiveness by reducing false alarms that occur when normal behavior for one individual appears suspicious based on general population patterns. The technology adapts to legitimate changes in user behavior while maintaining sensitivity to genuinely suspicious activities. Organizations implementing personalized risk profiling report substantial improvements in detection accuracy and reduced investigation costs due to fewer false positive alerts.

Continuous improvement through feedback loops ensures that fraud detection systems remain effective against evolving threats. Investigation feedback is systematically integrated into the scoring models, refining their accuracy based on real-world outcomes. When investigators determine that flagged activities were legitimate or identify missed fraud cases, this information is fed back into the system for model refinement. Scoring models are updated regularly to adapt to new fraud schemes and changing criminal tactics. This iterative improvement process ensures that the systems become more accurate over time, learning from both successes and failures. The feedback mechanism enables organizations to stay ahead of fraudsters who continuously adapt their methods, maintaining effective detection capabilities in a dynamic threat environment.

In conclusion, this micro course has demonstrated the critical importance of advanced fraud detection techniques in protecting insurance and financial services organizations. Anomaly detection provides the foundation for identifying unusual patterns that indicate potential fraudulent activities in real-time monitoring systems. Deep learning applications revolutionize image and document verification, enabling automated detection of sophisticated forgeries and manipulations. Cross-channel analysis integrates multiple data sources to provide comprehensive fraud insights that reveal coordinated criminal activities. Adaptive scoring systems dynamically adjust risk assessments based on evolving threat patterns, ensuring continuous protection against new fraud schemes. Together, these technologies create a comprehensive defense framework for modern financial services.

Chapter 7: AI-Powered Claims Processing

Automated First Notice of Loss represents a fundamental shift in how insurance claims begin. The AI-powered claim reporting system allows policyholders to initiate claims through digital platforms with unprecedented efficiency. The system automatically captures comprehensive claim data from multiple sources including mobile apps, web portals, and voice interactions. Advanced validation algorithms ensure data accuracy and completeness before processing begins. Once validated, claims are automatically forwarded to appropriate handlers based on predetermined criteria, eliminating manual routing delays. This automation minimizes human intervention in initial stages, reduces data entry errors, and significantly accelerates the claim initiation process, setting the foundation for faster overall resolution times.

The benefits of AI-driven FNOL extend to both insurers and their customers through measurable improvements. Processing time reduction is achieved through instant data capture and validation, eliminating traditional phone queues and manual data entry delays. Operational costs decrease substantially as automation reduces labor requirements and minimizes costly human errors that previously required correction. Customer experience enhancement occurs through 24/7 availability, real-time claim tracking, and immediate confirmation of claim receipt. Policyholders can report claims at their convenience using intuitive digital interfaces, receiving instant acknowledgment and case numbers. This transformation creates a win-

win scenario where insurers improve efficiency while customers enjoy faster, more convenient service delivery.

Despite significant benefits, AI-driven FNOL implementation faces several integration challenges requiring careful planning. Legacy system compatibility presents the most significant hurdle, as older insurance systems often use outdated technologies and incompatible data formats. Data security concerns intensify with digital transformation, requiring robust encryption and compliance with insurance regulations. User adoption challenges emerge as both customers and staff adapt to new processes, necessitating comprehensive training programs. Best practices for successful integration include phased implementation approaches that minimize disruption, establishing strong data governance frameworks, and providing continuous training and support. Organizations must also ensure seamless API connections between AI systems and existing infrastructure while maintaining data integrity throughout the transition process.

AI algorithms revolutionize claim prioritization through sophisticated machine learning analysis of multiple data points. These systems evaluate claim attributes including damage type, policy details, customer history, and external factors to determine appropriate categories and priority levels. Urgency and complexity categorization ensures that time-sensitive claims receive immediate attention while complex cases are routed to specialist adjusters. The algorithms continuously learn from historical claim outcomes, improving accuracy over time. Enhanced decision making occurs through predictive analytics that identify potential complications early in the process. This intelligent prioritization system ensures optimal resource allocation, reduces bottlenecks, and improves overall workflow management while maintaining consistent quality standards across all claim types.

Predictive models for severity assessment leverage diverse data sources to provide comprehensive risk evaluation capabilities. AI systems analyze claim history patterns, detailed customer profiles, and external factors such as weather conditions, traffic data, and regional risk indicators. These predictive severity models process vast amounts of structured and unstructured data to estimate potential claim costs and complexity levels. The models incorporate machine learning algorithms that identify subtle patterns and correlations invisible to human analysts. Cost and resource optimization benefits include more accurate reserves setting, improved vendor selection, and efficient adjuster assignment. This data-driven approach enables insurers to anticipate resource needs, allocate specialist expertise appropriately, and maintain optimal claim handling efficiency while controlling costs.

AI implementation in claim triage delivers substantial improvements in cycle times and resource allocation efficiency. Accurate claim triage ensures that each case receives appropriate attention level, with straightforward claims processed rapidly through automated workflows while complex cases receive dedicated specialist resources. Cycle time reduction occurs through elimination of manual sorting processes and immediate routing to qualified handlers. The system identifies cases requiring urgent attention, preventing delays that could escalate claim costs or customer dissatisfaction. Optimized resource allocation ensures senior adjusters focus on high-value or complex claims while routine cases proceed through streamlined automated processes. This intelligent distribution maximizes team productivity, reduces overall processing costs, and improves customer satisfaction through faster resolution times.

Computer vision technologies represent a breakthrough in automated damage assessment for insurance applications. Image recognition systems automatically identify and classify various types of damage in photographs submitted by policyholders or adjusters, dramatically reducing manual inspection requirements. Object detection algorithms pinpoint specific damaged components within images, enabling precise identification of affected parts, systems, or structures. Deep learning models continuously improve accuracy through exposure to vast datasets of damage images, learning to recognize increasingly subtle patterns and damage types. These advanced algorithms can distinguish between different severity levels, identify potential safety hazards, and even detect signs of pre-existing conditions. This technological advancement transforms subjective visual assessments into objective, consistent, and reproducible evaluations.

Image analysis capabilities extend beyond simple damage identification to comprehensive assessment and cost estimation. AI models evaluate damage severity levels with remarkable precision, analyzing factors such as impact area, depth of damage, and structural implications. Repair cost estimation algorithms combine damage assessment data with current market pricing, labor costs, and parts availability to provide accurate repair estimates. This automation significantly speeds up the claims processing and approval workflow. Fraud detection enhancement represents another critical capability, as AI systems can

63

identify inconsistencies in damage patterns, detect manipulated images, or flag suspicious claim characteristics. These systems compare submitted images against extensive databases of legitimate damage patterns, helping identify potentially fraudulent claims that require additional investigation.

Computer vision accuracy continues improving through advanced algorithms and enhanced data processing capabilities, though current limitations persist. Accuracy improvements result from better training datasets, more sophisticated neural networks, and improved image processing techniques that handle various lighting conditions, angles, and image qualities. Current limitations include challenges with poor image quality, unusual damage patterns, complex scenarios involving multiple impact types, and distinguishing between new and pre-existing damage. Weather conditions, image resolution, and photographer skill levels can affect system performance. Ongoing research focuses on addressing these limitations through improved algorithms, synthetic data generation, and multi-modal analysis combining images with other data sources. Future developments promise enhanced accuracy and broader application capabilities.

Virtual adjusters play increasingly important roles in modern claims resolution by handling routine tasks and supporting human adjusters. These AI-powered systems excel at claim documentation support, automatically organizing and categorizing submitted materials, extracting relevant information, and maintaining comprehensive case files. Status update automation ensures all stakeholders receive timely communications throughout the claims process without human intervention. Virtual adjusters provide preliminary assessments of straightforward claims, analyzing available information to determine coverage applicability, estimate damages, and identify cases requiring human expertise. This division of labor allows human adjusters to focus their expertise on complex, high-value, or disputed claims while ensuring routine cases progress efficiently. The result is improved overall capacity and faster processing times.

Insurance chatbots offer significant capabilities while facing specific limitations in customer service applications. Round-the-clock customer support availability ensures policyholders can access assistance regardless of time zones or business hours, improving overall service accessibility. These systems efficiently handle common inquiries such as policy information requests, claim status updates, and basic procedural questions. Chatbots guide users through claim submission processes, ensuring complete information collection and proper documentation. However, limitations emerge with complex queries requiring nuanced understanding, emotional support, or specialized expertise. These situations require escalation to human agents who can provide personalized attention and handle unique circumstances. The key to successful implementation lies in properly defining chatbot capabilities and ensuring seamless handoff processes to human agents when needed.

AI implementation delivers measurable improvements in both customer experience and operational efficiency across insurance organizations. Enhanced customer experience results from faster response times, consistent service quality, and personalized interactions based on customer history and preferences. AI systems provide immediate acknowledgment of communications, accurate information delivery, and proactive updates throughout the claims process. Customers appreciate the convenience of 24/7 availability and self-service options for routine tasks. Operational efficiency improvements include significant reductions in call center volume as customers find answers through automated channels. Workflow streamlining occurs as AI handles routine tasks, allowing human agents to focus on complex issues requiring empathy and specialized knowledge. This optimization reduces operational costs while improving service quality and customer satisfaction scores.

AI transformation in insurance claims processing delivers comprehensive benefits across multiple operational dimensions. Automation in claims workflows reduces manual tasks substantially, eliminating repetitive data entry, routing decisions, and status updates while accelerating overall processing times. Improved accuracy results from consistent AI analysis that eliminates human error variability and applies standardized evaluation criteria across all claims. Enhanced customer experience emerges through faster responses, consistent service quality, and convenient self-service options that meet modern consumer expectations. The competitive advantage gained through AI adoption enables insurers to increase operational efficiency, reduce costs, and improve customer satisfaction simultaneously. Organizations implementing these technologies position themselves as industry leaders while building sustainable competitive advantages in an increasingly digital marketplace.

Chapter 8: Enhancing Customer Experience

Conversational AI represents a fundamental shift in customer service delivery, encompassing chatbots, virtual assistants, and sophisticated natural language processing technologies. These systems enable truly human-like interactions that feel natural and intuitive to customers. The key benefits include enhanced accessibility for users with different needs and preferences, dramatically accelerated service response times compared to traditional support channels, and consistent interaction quality regardless of time or day. Unlike human agents who may have varying expertise levels or emotional states, AI maintains uniform service standards while handling multiple conversations simultaneously, creating scalable customer service solutions.

Automated customer interactions deliver measurable operational improvements across multiple dimensions. Reduced wait times occur because AI provides instant responses and quick solutions without queue delays. The 24/7 availability ensures customers receive assistance whenever they need it, eliminating the frustration of business hours limitations. Lower operational costs result from minimizing the need for extensive human staff in routine, repetitive tasks, allowing organizations to reallocate resources to complex problem-solving. Enhanced customer satisfaction emerges when instant automated responses handle simple queries effectively, freeing human agents to focus on sophisticated issues that require empathy, creativity, and advanced problem-solving skills.

Successful AI implementation requires strategic planning and adherence to proven best practices. Clear use cases must be identified from the start to ensure AI solutions target relevant problems effectively rather than implementing technology for its own sake. Seamless integration with existing systems is essential for smooth AI deployment and proper data flow between platforms. Continuous AI training helps models adapt to evolving data patterns and improve performance consistently over time. Customer feedback monitoring provides crucial insights for refining AI interactions and enhancing user satisfaction. Regular assessment and adjustment ensure the AI system remains aligned with business objectives and customer expectations.

Data-driven personalization transforms how insurance and financial services engage with customers by leveraging comprehensive customer information. AI systems analyze demographics, behavioral patterns, and stated preferences to create detailed customer profiles that inform product recommendations. This approach moves beyond one-size-fits-all solutions to deliver truly customized experiences. Customer data utilization enables organizations to understand individual risk profiles, financial goals, and life circumstances. Personalized policy options maximize client value by presenting relevant choices that align with specific needs, increasing customer satisfaction and purchase likelihood while reducing decision fatigue from overwhelming product catalogs.

AI algorithms excel at identifying complex patterns in customer data to understand profiles and preferences with remarkable accuracy. Data pattern analysis uses machine learning models to process vast amounts of information, revealing insights that human analysts might miss. These systems recognize subtle correlations between demographic factors, behavioral indicators, and product preferences. Personalized policy recommendations emerge from AI analysis of individual risk profiles and financial goals, ensuring suggestions align with customer circumstances. Enhanced decision accuracy results from AI-driven insights that consider multiple variables simultaneously, leading to better outcomes and greater customer satisfaction through more relevant, timely recommendations.

Real-world implementations demonstrate AI's measurable impact on business performance and customer engagement. Personalized AI recommendations significantly enhance policy uptake rates and customer engagement metrics compared to generic marketing approaches. Case studies consistently show increased customer retention through tailored AI solutions that make customers feel understood and valued. Revenue growth impact becomes apparent as AI-based personalized strategies drive higher conversion rates and customer lifetime value. Organizations implementing these technologies report improved customer acquisition costs, reduced churn rates, and increased average transaction values. The data clearly supports AI personalization as a competitive advantage that delivers tangible business results.

Real-time sentiment detection employs sophisticated techniques to understand customer emotions during interactions. Natural Language Processing analyzes text input to detect positive, negative, or neutral sentiments accurately and instantaneously. Machine learning algorithms classify emotions from both voice tonality and text inputs, enabling rapid sentiment detection during live interactions. Customer mood detection provides valuable insights that help gauge emotional states during support calls, allowing for more responsive and appropriate service delivery. These technologies work together to create a comprehensive understanding of customer emotional states, enabling support teams to respond with greater empathy and effectiveness.

Integration of sentiment analysis with existing support platforms creates powerful capabilities for enhanced customer service. Sentiment detection integration embeds emotional analysis directly into CRM and helpdesk systems, providing real-time emotional monitoring without disrupting established workflows. Emotion alerts help prioritize urgent customer cases, ensuring frustrated or upset customers receive immediate attention and preventing escalation. Guided response strategies emerge from sentiment insights, helping support agents craft appropriate and empathetic responses based on detected emotional states. This integration transforms customer support from reactive problem-solving to proactive emotional management, improving both resolution outcomes and customer satisfaction.

Proactive sentiment management significantly improves resolution efficiency and customer satisfaction outcomes. Addressing negative customer sentiments early in interactions helps prevent conflicts and escalations that consume additional resources and damage relationships. Faster resolution times result from prioritizing emotionally charged cases and providing agents with context about customer emotional states. This approach reduces average handling times while improving first-call resolution rates. Boosted customer satisfaction emerges from faster resolutions and reduced conflicts, as customers feel heard and understood rather than transferred between departments. The combination creates a positive feedback loop where better emotional management leads to improved outcomes and stronger customer relationships.

AI-powered FAQ systems represent a significant advancement over static knowledge bases through continuous adaptation and learning. These systems learn from customer interactions, analyzing which questions are asked most frequently and how users phrase their inquiries. AI learning from interactions enables the system to provide increasingly accurate answers over time. Continuous response improvement occurs as the system updates and refines responses based on customer feedback and interaction patterns. Enhanced customer experience results from adaptive FAQs that ensure customers receive timely, relevant information tailored to their specific needs. This dynamic approach keeps information current and useful rather than outdated and frustrating.

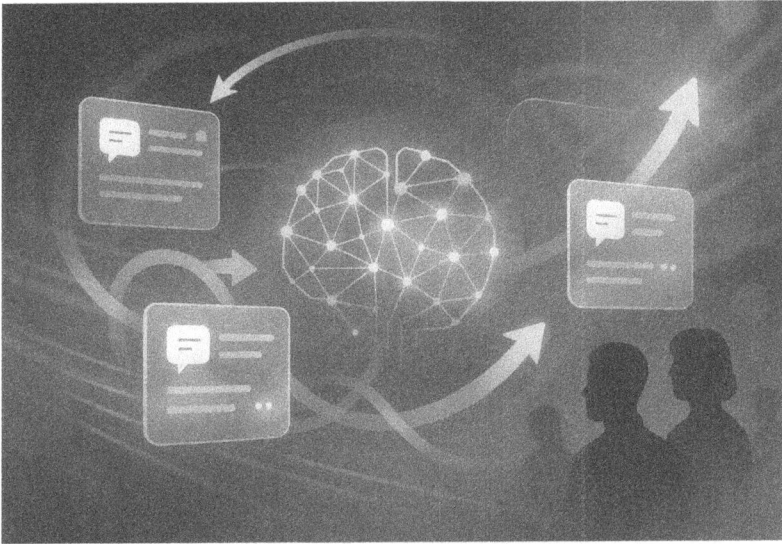

Knowledge agents handle complex queries that exceed typical FAQ capabilities through advanced understanding and comprehensive response generation. These agents interpret complex questions by analyzing intricate and nuanced queries that require contextual understanding rather than simple keyword matching. Multi-source information retrieval allows agents to gather relevant information from diverse databases, documents, and knowledge repositories efficiently. Detailed answer generation provides comprehensive responses that go beyond typical FAQ entries, offering thorough explanations with relevant context. This capability bridges the gap between simple automated responses and human expertise, providing sophisticated assistance while maintaining the efficiency benefits of AI-powered support systems.

Continuous improvement ensures knowledge bases remain current and effective through systematic updating processes. Ongoing data collection gathers information from customer interactions, support tickets, and emerging issues to keep knowledge bases relevant over time. AI training enhancement involves regular model updates that improve accuracy and responsiveness to new information and changing customer needs. Dynamic adaptation enables knowledge bases to evolve with changing customer requirements and product updates automatically. This approach ensures that AI systems don't become stale or outdated but instead grow more capable and useful over time, maintaining their value proposition as customer needs and business offerings evolve.

AI technologies create transformative improvements in customer experience through multiple interconnected benefits. AI enhances customer experience by enabling efficient self-service options and personalized recommendations that improve satisfaction and engagement levels significantly. Empathetic and intelligent support emerges from AI systems that provide understanding responses and comprehensive knowledge management, making customer service more effective and responsive to individual needs. Competitive advantage and loyalty result from adopting these advanced technologies, fostering long-term customer relationships while helping businesses stay ahead in increasingly competitive markets. Organizations implementing comprehensive AI customer experience strategies position themselves for sustained growth and customer retention.

Chapter 9: Policy Servicing and Lifecycle Management

Policy servicing encompasses three critical areas that form the foundation of effective insurance operations. The policy lifecycle activities include issuing new policies, modifying existing coverage, processing renewals, and managing terminations efficiently. Customer satisfaction directly correlates with how well these processes function, as timely and accurate services create positive experiences that build trust and loyalty. Regulatory compliance ensures all policy servicing activities adhere to legal and insurance industry requirements, protecting both the insurer and policyholder. When these three elements work harmoniously, insurance companies can deliver superior service while maintaining operational excellence and meeting all regulatory obligations.

The insurance policy lifecycle consists of five distinct stages, each requiring specific attention and processes. The application stage initiates the relationship, where detailed client information and coverage needs are carefully collected and documented. During underwriting, risk assessment professionals evaluate applications to determine appropriate policy terms, coverage limits, and premium calculations based on individual risk profiles. Policy issuance follows successful underwriting, involving formal documentation preparation and delivery to policyholders. The servicing and renewal phase manages ongoing policy updates, premium payments, and continuous coverage maintenance. Finally, claim settlement processes handle policy-based claims according to established terms, completing the comprehensive insurance lifecycle and ensuring policyholder protection.

Technology plays a transformative role in modern lifecycle management through three key mechanisms. Process acceleration utilizes AI and automation to significantly speed up traditionally time-consuming lifecycle management processes, improving overall efficiency while reducing delays that can frustrate customers and impact business operations. Error reduction represents another crucial benefit, as technology minimizes human errors that can occur during manual data processing, policy updates, and documentation management, ensuring more accurate and reliable lifecycle management operations. Proactive customer engagement becomes possible through technology platforms that enable personalized, timely interactions throughout the entire policy lifecycle, creating opportunities for better communication and enhanced customer relationships.

Automated identification of renewal dates represents a fundamental advancement in policy management efficiency. AI-driven data analysis enables systems to process vast amounts of policy information automatically, identifying upcoming renewal dates with precision and consistency that surpasses manual tracking methods. Automated renewal tracking eliminates human oversight risks by maintaining continuous monitoring of all policies within the portfolio, ensuring no renewal opportunities are missed due to administrative oversights. Timely notifications generated by AI systems provide consistent, reliable alerts well in advance of renewal deadlines, giving both insurance teams and customers adequate time to review coverage options, make necessary adjustments, and complete renewal processes without rushed decisions or lapses in coverage.

Personalized customer notifications leverage artificial intelligence to create more effective communication strategies that resonate with individual policyholders. AI-driven personalization analyzes customer profiles, communication preferences, policy history, and behavioral patterns to customize renewal reminders that match each individual's specific needs and preferences, making communications more relevant and actionable. Enhanced customer engagement results from these tailored communications, as customers respond more positively to messages that feel personally relevant rather than generic mass communications. This approach significantly increases response rates, reduces customer service inquiries, and improves overall satisfaction levels by demonstrating that the insurance company understands and values each customer's unique circumstances and preferences.

The impact on retention and customer experience demonstrates clear business value from AI-driven renewal systems. AI-driven renewal proactiveness enables insurance companies to engage customers well before policy expiry dates, providing multiple touchpoints for communication and discussion about coverage options. This early engagement significantly increases policy retention rates by ensuring customers feel valued and informed throughout the renewal process. Improved customer satisfaction emerges from relevant, personalized AI interactions that make customers feel understood and appreciated rather than treated as anonymous policy numbers. These enhanced experiences create positive emotional connections with the insurance brand, leading to higher satisfaction scores, increased loyalty, and greater likelihood of policy renewals and additional product purchases.

AI-powered summarization of policy documents addresses one of the most time-consuming aspects of insurance operations. Natural Language Processing technology efficiently processes complex text data within lengthy policy documents, extracting essential information that would traditionally require hours of manual review. This automated approach ensures consistency in information extraction while dramatically reducing processing time. Concise policy summaries generated by AI highlight the most critical policy points, terms, conditions, and coverage details in easily digestible formats that benefit both clients and insurance agents. These summaries enable quick comprehension of complex policy language, facilitate faster decision-making, and improve communication between insurance professionals and policyholders by presenting information in clear, accessible language.

Automated translation for multilingual clients removes language barriers that can complicate insurance operations in diverse markets. AI-powered language translation systems enable fast, accurate translation across multiple languages, supporting global communication needs without requiring extensive human translation resources. These systems maintain consistency in terminology and ensure that complex insurance concepts are properly conveyed across different languages and cultural contexts. Breaking language barriers through automated translation significantly improves services for international clients by making policy information, communications, and documentation accessible regardless of language preferences. This capability expands market reach, improves customer satisfaction among diverse populations, and enables insurance companies to serve global markets more effectively while maintaining service quality standards.

The benefits for global policy servicing demonstrate the transformative impact of AI-powered document processing technologies. Enhanced accessibility ensures that policies become available across different languages and regions, broadening global reach and making insurance products accessible to diverse populations regardless of linguistic backgrounds. Reduced processing time results from summarization tools that speed up policy review processes by condensing complex information efficiently and accurately, enabling faster decision-making and improved operational efficiency. Improved compliance emerges as automated technologies help ensure that policies consistently meet international standards and regulatory requirements across different jurisdictions, reducing compliance risks and maintaining regulatory adherence while operating in multiple markets with varying requirements.

Extracting key terms and conditions using NLP revolutionizes how insurance professionals analyze policy documents. NLP for policy analysis employs sophisticated algorithms to automatically analyze policy texts, identifying and extracting important terms, conditions, exclusions, and coverage details that traditionally required manual review by experienced professionals. Enhancing clarity becomes possible as extracted key terms improve understanding and readability of complex policy documents, making them more accessible to both insurance professionals and policyholders. Risk assessment support emerges through the identification of essential clauses and conditions that directly impact risk evaluation, enabling more accurate risk assessment processes and better-informed underwriting decisions that protect both insurers and policyholders from unexpected coverage gaps or disputes.

Ensuring regulatory compliance through automated analysis represents a significant advancement in insurance operations management. NLP-powered compliance checks automate the traditionally manual process of reviewing policies against regulatory requirements, significantly reducing reliance on time-consuming human reviews while improving accuracy and consistency. These systems can simultaneously check policies against multiple regulatory frameworks and identify potential compliance issues before they become problematic. Early detection of issues allows automated systems to quickly identify potential regulatory problems before they escalate into serious compliance violations, enabling proactive resolution and preventing costly regulatory penalties. This approach protects insurance companies from compliance risks while ensuring that all policies meet current regulatory standards and requirements.

Reducing human error and improving accuracy addresses fundamental challenges in insurance operations where precision is critical. Minimizing manual errors becomes achievable as AI significantly reduces mistakes that commonly occur during manual data interpretation, policy analysis, and document processing activities. These automated systems maintain consistent performance standards without the variability inherent in human processing. Enhancing process accuracy results from AI implementation that improves overall reliability and precision of policy servicing workflows, ensuring that decisions are based on complete, accurate information rather than potentially flawed manual analysis. This increased accuracy reduces disputes, improves customer satisfaction, and protects insurance companies from errors that could lead to coverage disputes or regulatory issues.

AI-driven workflow optimization transforms how insurance operations manage workloads and prioritize tasks for maximum efficiency. Workload analysis enables AI systems to evaluate current workloads across the organization, identifying tasks that require immediate attention and those that can be scheduled for optimal resource utilization. This analysis considers factors such as complexity, urgency, resource requirements, and deadlines to create comprehensive workload pictures. Prioritization of tasks ensures that AI systems can intelligently rank servicing tasks based on multiple criteria including customer impact, regulatory requirements, and business objectives, ensuring timely and effective completion of critical work. Workflow streamlining optimizes operational processes by removing inefficiencies, reducing redundant steps, and creating smoother pathways for task completion that minimize delays and maximize productivity.

Dynamic assignment of servicing tasks creates more efficient and effective resource utilization across insurance operations. Task allocation by skills enables intelligent systems to evaluate individual agent capabilities, experience levels, and expertise areas to assign tasks where knowledge and skills most closely match specific task requirements, improving both quality and efficiency outcomes.

Availability consideration ensures that systems monitor agent schedules, current workloads, and capacity to optimize task distribution while avoiding overloading any individual team members. Workload balancing dynamically distributes tasks across available resources to maintain even workloads that prevent burnout, improve job satisfaction, and maintain consistent service quality. This balanced approach ensures optimal resource utilization while supporting employee well-being and maintaining high performance standards.

Enhancing operational efficiency and productivity represents the culmination of intelligent routing and task assignment systems working together seamlessly. Task routing optimization ensures that AI systems direct work through the most efficient pathways, creating smooth workflows that minimize handoffs, reduce processing time, and accelerate task completion while maintaining quality standards. Bottleneck reduction becomes possible as AI identifies and addresses workflow constraints by streamlining processes and balancing workloads more effectively, preventing delays that can cascade through operations. Improved operational efficiency emerges as the overall result of AI-driven optimizations that enhance productivity, reduce costs, improve customer satisfaction, and enable insurance operations to handle larger volumes of work with existing resources while maintaining or improving service quality standards.

In conclusion, artificial intelligence is fundamentally transforming policy servicing by automating complex processes and significantly increasing operational efficiency across all areas of insurance operations. These AI implementations reduce manual work, minimize errors, and enable more strategic resource allocation. Natural language processing improvements are revolutionizing customer experience by enabling better communication, faster response times, and more accurate information processing that directly benefits client interactions. These technologies position modern insurers to effectively address contemporary challenges including increasing customer expectations, regulatory complexity, and competitive pressures while meeting evolving client needs. The integration of AI and NLP creates sustainable competitive advantages that improve both operational performance and customer satisfaction in today's dynamic insurance marketplace.

Chapter 10: Generative AI and New Product Development

We begin our exploration with Generative AI and New Product Development. This foundational section will establish the technological framework and demonstrate how AI is fundamentally changing traditional product development approaches in the insurance industry. We'll examine the core technologies, their applications, and the transformative benefits they bring to insurers and customers alike.

Generative AI represents a paradigm shift in insurance product development. At its core, generative AI uses advanced algorithms to create entirely new content and solutions across various domains. Key technologies include sophisticated language models and generative adversarial networks that enable unprecedented content generation capabilities. In insurance specifically, generative AI enhances creative problem-solving by automating complex product design processes, analyzing vast datasets to identify market opportunities, and generating innovative solutions that traditional methods might miss entirely.

The benefits of AI in product innovation are transformative across multiple dimensions. AI dramatically accelerates development cycles by generating ideas automatically and eliminating repetitive manual tasks that traditionally slow product launches. Perhaps more importantly, AI excels at predicting customer needs with remarkable accuracy, enabling insurers to tailor products for optimal market fit and enhanced customer satisfaction. Additionally, AI significantly improves product personalization and scalability, allowing insurance companies to offer customized solutions at unprecedented scale while maintaining competitive pricing and market positioning.

While AI adoption offers tremendous benefits, several critical challenges must be addressed for successful implementation. Data quality issues represent the most fundamental concern, as poor data undermines AI effectiveness and leads to flawed decision-making in product development. Regulatory compliance presents another significant hurdle, requiring insurers to navigate complex insurance regulations while implementing new AI technologies legally. Ethical considerations demand careful attention to ensure responsible AI use that protects consumer interests. Finally, integration complexity poses substantial technical and operational challenges when incorporating AI systems into existing insurance infrastructure and workflows.

Moving to our second major topic, AI for Microinsurance Design represents a critical application area where technology meets social impact. This section explores how AI enables insurers to develop affordable, accessible insurance products for low-income populations and underserved markets, fundamentally expanding insurance accessibility worldwide.

Understanding microinsurance needs requires deep insight into the unique challenges facing low-income populations and their specific risk preferences. Traditional market research methods often fall short when analyzing these underserved segments. AI data analysis fills this gap by processing vast amounts of demographic and behavioral data to scale understanding of microinsurance needs effectively. Machine learning algorithms can identify patterns in spending behavior, risk exposure, and insurance preferences that would be impossible to detect through conventional analysis, enabling insurers to develop truly relevant products for these markets.

AI-driven customer segmentation revolutionizes how insurers identify and serve target markets in microinsurance. Advanced analytics enable precise segmentation of customers by analyzing complex risk profiles and insurance needs with unprecedented accuracy. This capability allows insurers to move beyond broad demographic categories to understand nuanced customer requirements. Most importantly, this technology enables insurers to focus specifically on underserved and relevant customer segments, improving both service delivery and outreach effectiveness. The result is better-targeted products that genuinely meet the needs of previously overlooked market segments.

Personalized microinsurance offerings via AI represent the next evolution in affordable insurance products. AI-driven customization enables the creation of tailored microinsurance products that meet specific customer needs more effectively than one-size-fits-all approaches. The technology optimally aligns insurance coverage and pricing with individual personal circumstances, dramatically improving value proposition and customer satisfaction. This personalization leads to enhanced customer adoption rates by addressing unique customer requirements that traditional products often overlook, ultimately expanding insurance access to populations who previously found insurance products irrelevant or unaffordable.

Our third focus area explores On-Demand Insurance Models, representing a fundamental shift toward flexible, event-specific coverage. This emerging model leverages technology to provide short-term insurance solutions that align perfectly with modern consumer lifestyles and expectations for immediate, customizable services.

On-demand insurance fundamentally redefines traditional coverage models by providing short-term, event-specific coverage tailored precisely to individual needs and timing requirements. Rising digital adoption serves as a key enabler, making insurance policies more accessible and manageable through mobile platforms and apps. IoT devices play a crucial supporting role by enabling real-time monitoring and enhancing both personalized and event-specific insurance coverage capabilities. Modern consumer lifestyles increasingly drive demand for flexible and highly customizable insurance options available on short notice, reflecting broader trends toward on-demand services across industries.

Chapter 10: Generative AI and New Product Development

AI transforms underwriting and risk assessment by enabling rapid analysis of large data volumes, facilitating swift identification of risk factors that traditionally required extensive manual review. This technological capability dramatically improves accuracy in risk evaluation by utilizing advanced algorithms and real-time data streams that capture dynamic risk conditions. Perhaps most importantly, AI enables dynamic policy issuance through real-time underwriting decisions, significantly reducing processing times and streamlining the entire policy issuance process. This transformation allows insurers to offer instant coverage decisions that match consumer expectations for immediate service.

Customer experience enhancements through automation represent a critical competitive advantage in on-demand insurance. AI automation delivers instant quotes, dramatically speeding up customer decision-making processes and reducing friction in the purchase journey. Automated systems also simplify claims processing, reducing wait times and significantly enhancing overall customer satisfaction with the insurance experience. Additionally, AI provides highly personalized recommendations based on individual risk profiles and preferences, improving customer engagement levels and increasing retention rates through more relevant product suggestions and communications tailored to each customer's specific circumstances.

Our fourth major topic examines Generative AI for Policy Drafting, exploring how artificial intelligence is revolutionizing the creation, customization, and management of insurance policy documents. This application demonstrates AI's potential to enhance both efficiency and accuracy in critical insurance documentation processes.

Automating policy documentation with AI transforms traditional document creation processes through sophisticated AI-powered drafting capabilities. Generative AI efficiently creates detailed policy documents by intelligently combining client-specific data with relevant regulatory requirements and industry standards. This automation significantly reduces human errors that commonly occur in manual policy documentation processes, ensuring greater accuracy and consistency. Most importantly, AI dramatically accelerates policy document completion, improving overall turnaround times and enabling insurers to respond more quickly to customer needs while maintaining high-quality documentation standards throughout the process.

AI significantly improves clarity and compliance in policy wording through multiple mechanisms. The technology aids in creating clear, understandable policy language that reduces customer confusion and minimizes potential misinterpretation of coverage terms. AI maintains uniform terminology and consistent style across all policy documents, ensuring coherent communication throughout an insurer's product portfolio. Additionally, AI systems verify that policy wording complies with relevant laws and regulations, substantially reducing risks of non-compliance issues. Finally, AI identifies and minimizes ambiguous language that could lead to disputes, clarifying policy intent and protecting both insurers and policyholders.

Customization of policies to client needs represents a major advancement in personalized insurance service delivery. AI conducts detailed client profile analysis, examining comprehensive client profiles and preferences to understand unique individual needs and specific circumstances. This analysis enables tailored policy generation, where AI creates customized policy terms and conditions specifically designed for individual client requirements rather than generic templates. The result is enhanced customer satisfaction, as personalized policies increase relevance and significantly improve overall customer satisfaction and loyalty by demonstrating genuine understanding of individual customer needs.

Our final topic focuses on the Use of AI for Pricing Simulations, examining how artificial intelligence revolutionizes pricing strategies through sophisticated modeling, scenario analysis, and real-time adjustments. This critical application area directly impacts insurer profitability and market competitiveness.

AI-powered pricing models and algorithms represent a fundamental advancement in insurance pricing methodology. Machine learning algorithms process both historical and real-time data streams to understand complex market trends and risk patterns more effectively than traditional statistical methods. These sophisticated pricing models utilize AI insights to accurately reflect true underlying risks and respond dynamically to changing market conditions. The result is pricing that more precisely matches actual risk exposure, improving both profitability for insurers and fairness for customers through more accurate risk-based pricing structures.

Scenario simulation for risk and profitability enables comprehensive strategic planning through AI-powered risk scenario simulation capabilities. AI systems can simulate various insurance risk scenarios with remarkable accuracy, enabling insurers to predict potential outcomes across multiple market conditions and risk environments. These simulated scenarios directly help insurers refine their pricing strategies to enhance overall profitability while maintaining financial stability. This capability allows insurers to test pricing strategies against various market conditions before implementation, reducing the risk of pricing errors and improving long-term financial performance.

Dynamic adjustments based on real-time data represent the cutting edge of responsive pricing strategies. AI systems utilize continuous data feeds to monitor market conditions and identify emerging risks instantly as they develop. This real-time monitoring enables dynamic pricing adjustments, where pricing is continuously optimized by AI to maintain the optimal balance between risk exposure and reward potential. These systems effectively respond to market shifts by automatically balancing potential risks with expected rewards, ensuring that pricing remains competitive while protecting profitability across changing market conditions and risk environments.

In conclusion, generative AI's impact on insurance product development is both profound and far-reaching. AI drives unprecedented innovation and efficiency throughout insurance product development processes, fundamentally reshaping traditional approaches and methodologies. The personalization benefits cannot be overstated - AI enables truly personalized insurance products tailored to individual customer needs, resulting in significantly better customer satisfaction and market fit. Most importantly, adopting generative AI provides insurers with a crucial competitive advantage, helping them stay relevant and competitive by effectively meeting rapidly evolving market demands and customer expectations in an increasingly digital insurance landscape.

Chapter 11: AI in Distribution and Sales

Artificial intelligence transforms lead management by analyzing vast datasets to identify the most promising sales opportunities. The system processes historical sales data, customer behavioral patterns, website interactions, and demographic information to create comprehensive lead profiles. Machine learning algorithms then rank these leads based on conversion probability, allowing sales teams to focus their efforts strategically. This data-driven approach eliminates guesswork and ensures resources are allocated to prospects with the highest potential value. The result is improved sales efficiency, shortened sales cycles, and better resource utilization across the entire sales organization.

Effective lead scoring relies on diverse data sources to build accurate predictive models. CRM records provide historical interaction data, while website analytics reveal engagement patterns and behavioral indicators. Social media signals offer additional insights into prospect interests and purchase intent. Advanced predictive modeling techniques, including logistic regression for probability calculations and decision trees for rule-based scoring, form the foundation of intelligent lead assessment. Neural networks take this further by identifying complex, non-linear patterns in customer data that traditional methods might miss, enabling more sophisticated forecasting of lead quality and conversion likelihood.

AI-powered lead scoring delivers measurable benefits across sales operations. Intelligent prioritization reduces time wasted on low-quality leads, allowing sales teams to focus on prospects most likely to convert. This focused approach accelerates deal closure timelines by ensuring high-value opportunities receive appropriate attention and resources. Pipeline management becomes more efficient through better visibility into lead quality and progression likelihood. The cumulative effect is significantly higher conversion rates, as sales efforts are concentrated on the most promising opportunities, ultimately driving improved revenue performance and sales team productivity.

AI copilots represent a new paradigm in sales support, functioning as intelligent assistants embedded directly within sales platforms and workflows. These systems enhance agent productivity by automating routine administrative tasks, particularly repetitive data entry that traditionally consumes valuable selling time. Beyond automation, copilots provide real-time insights and recommendations, analyzing customer data and interaction history to suggest optimal next actions. This intelligent assistance enables sales representatives to focus on high-value activities like relationship building and strategic selling, while ensuring all customer interactions are informed by comprehensive data analysis and best-practice recommendations.

Real-time assistance capabilities transform customer interactions by providing instant access to critical information. AI copilots deliver comprehensive product details, specifications, and availability data during live customer conversations, eliminating delays and improving response accuracy. Dynamic pricing information and competitive analysis help agents position offers effectively and respond to price objections confidently. Access to current competitor data enables informed discussions about value propositions and market positioning.

This immediate knowledge availability enhances customer experience through faster, more accurate responses while building agent confidence and credibility during sales conversations.

AI automation fundamentally changes how sales agents spend their time and energy. By handling routine administrative tasks, artificial intelligence frees representatives to engage in meaningful customer relationship building and strategic selling activities. Agents can dedicate more time to understanding customer needs, developing personalized solutions, and nurturing long-term partnerships. This shift from transactional to relationship-focused selling improves customer satisfaction through more attentive, personalized service. Enhanced productivity results from eliminating administrative bottlenecks, allowing agents to handle more prospects effectively while maintaining high service quality and building stronger customer connections.

AI-powered churn prediction transforms customer retention by identifying at-risk accounts before disengagement occurs. Advanced algorithms analyze transactional patterns, engagement metrics, support interactions, and usage trends to detect early warning signals of potential customer departure. This predictive capability enables businesses to implement proactive retention strategies, addressing customer concerns and dissatisfaction before they result in churn. Early risk detection allows for timely intervention through personalized outreach, special offers, or enhanced support services. The proactive approach is significantly more cost-effective than reactive customer acquisition, protecting revenue streams and maintaining valuable customer relationships.

Personalized retention strategies leverage AI-driven customer insights to create targeted intervention approaches. Machine learning algorithms analyze individual customer preferences, behaviors, and interaction history to develop customized retention offers that resonate with specific customer needs and motivations. Communication strategies are optimized based on customer channel preferences and response patterns, ensuring messages reach customers through their preferred touchpoints. Tailored offers, whether pricing adjustments, service upgrades, or exclusive benefits, address individual customer value perceptions and pain points. This personalized approach significantly improves retention campaign effectiveness compared to generic, one-size-fits-all strategies.

Real-world implementations demonstrate measurable churn reduction through AI-powered strategies. Organizations implementing predictive churn models report significant improvements in customer retention rates and reduced acquisition costs. Case studies show successful deployment of AI models that accurately identify at-risk customers months before potential departure, enabling timely intervention. Customized retention strategies, informed by AI insights, achieve higher success rates than traditional approaches. The combined methodology results in quantifiable decreases in customer loss, improved customer lifetime value, and enhanced long-term sales performance, demonstrating clear return on investment from AI-powered retention initiatives.

Chapter 11: AI in Distribution and Sales

Automated content creation revolutionizes marketing campaign development by generating diverse content types including blog posts, social media updates, email campaigns, and advertisements. AI tools analyze brand guidelines, target audience characteristics, and campaign objectives to produce relevant, on-brand content at scale. The system ensures consistency across all marketing channels while adapting tone and messaging for different audience segments. Content creation that traditionally required hours or days can be completed in minutes, dramatically reducing campaign development timelines. This automation allows marketing teams to focus on strategy, analysis, and optimization rather than manual content production tasks.

Targeted messaging capabilities enable unprecedented personalization in marketing communications. AI analyzes customer segments based on demographics, behavioral patterns, purchase history, and engagement preferences to create precise audience targeting. Personalized messages increase relevance and engagement by addressing specific customer interests, needs, and pain points. Dynamic content adaptation ensures each customer receives communications tailored to their preferences and stage in the buying journey. This personalization approach significantly improves conversion potential by delivering the right message to the right customer at the optimal time, maximizing campaign effectiveness and customer response rates.

Measuring AI-generated content impact requires comprehensive analytics across multiple performance indicators. Engagement rate tracking monitors audience interaction levels, including clicks, shares, comments, and time spent with content, providing insights into content effectiveness and audience resonance. Lead generation analysis evaluates how AI-driven content attracts and captures potential customer interest, measuring conversion from content consumption to qualified leads. Sales conversion measurement tracks the ultimate success metric: turning prospects into paying customers through AI-generated content touchpoints. Continuous optimization uses this data to refine AI algorithms and content strategies for improved performance.

In conclusion, this chapter has demonstrated how AI transforms four critical areas of sales and distribution. Enhanced lead targeting through intelligent scoring systems enables more effective prospect prioritization and resource allocation. AI copilots empower sales agents with real-time insights and automation, improving productivity and customer engagement quality. Predictive churn models enable proactive retention strategies that protect revenue and strengthen customer relationships. Personalized marketing through AI-driven content generation creates more relevant, effective campaigns. Together, these innovations represent a comprehensive transformation of traditional sales processes, driving measurable improvements in efficiency, effectiveness, and business results.

Chapter 12: Strategic AI Governance

Successful AI governance begins with developing organizational capabilities through strategic workforce development. Training existing staff provides essential skills for technology evolution, while recruiting specialized AI professionals strengthens knowledge bases in insurance-specific applications. Fostering continuous learning ensures teams stay current with rapidly advancing AI technologies. This three-pronged approach creates a sustainable foundation for AI adoption, combining internal development with external expertise to build comprehensive organizational competency in artificial intelligence applications.

Data literacy forms the cornerstone of effective AI implementation across insurance organizations. When teams understand data's value for decision-making and collaboration, they become more effective partners in AI initiatives. Cross-team collaboration improves significantly when all departments speak the same data language, leading to enhanced operational efficiency. Data literacy bridges the gap between technical AI teams and business units, ensuring that insights generated by AI systems are properly understood, interpreted, and acted upon throughout the organization.

Integrating AI into core insurance processes transforms operational efficiency and customer experience. AI in underwriting automates risk assessment and policy approval, dramatically improving both speed and accuracy. Claims processing becomes faster and more precise through AI-powered analysis and validation systems. Customer service enhancement through AI-driven chatbots and virtual assistants provides personalized, immediate responses. These applications demonstrate AI's practical value while maintaining the human oversight necessary for complex insurance decisions and relationship management.

Effective AI governance requires systematic identification and mitigation of technology-related risks. Model bias risks threaten fairness and reliability in AI-driven insurance decisions, requiring proactive detection and correction mechanisms. Data quality issues directly impact model accuracy and must be addressed through rigorous data management processes. Operational vulnerabilities expose AI systems to security threats and system failures, necessitating comprehensive protection strategies. Understanding and addressing these risks ensures AI deployment serves business objectives while maintaining customer trust and regulatory compliance.

Explainable AI methodologies are essential for stakeholder confidence and regulatory compliance. Frameworks that provide clear insights into AI decision-making processes make complex outputs understandable to all stakeholders. When stakeholders can interpret and validate AI outcomes, trust increases significantly across customer, regulatory, and internal audiences. Explainability also supports accountability by helping organizations meet evolving regulatory and ethical standards. This transparency becomes increasingly important as AI systems handle more critical insurance decisions affecting customer outcomes.

Transparency serves as the foundation for stakeholder confidence in AI-powered insurance operations. Clear explanations of AI model workings help build trust among customers, regulators, and internal teams by demystifying complex algorithmic processes. When stakeholders understand how AI systems make decisions, confidence increases in both the technology and the organization deploying it.

Chapter 12: Strategic AI Governance

Transparent communication about AI capabilities, limitations, and decision processes creates an environment of trust that supports broader AI adoption while maintaining necessary oversight and accountability.

Addressing bias in AI decision-making is fundamental to fair and ethical insurance operations. Bias detection systems identify unfair patterns in AI models that could lead to discriminatory outcomes, protecting both customers and organizations from legal and reputational risks. Mitigation strategies apply proven techniques to reduce bias, ensuring AI systems treat all users fairly and ethically. These efforts support legal compliance while promoting equal treatment of customers, creating insurance processes that are both technologically advanced and socially responsible.

Ethical standards in AI deployment require clear guidelines and governance frameworks that respect human rights and societal norms.

Chapter 12: Strategic AI Governance

Organizations must establish comprehensive ethical guidelines that govern AI development and implementation decisions. Governance frameworks support accountable and transparent AI deployment across all organizational levels and functions. Aligning AI use with organizational values builds public trust and meets evolving societal expectations. This alignment ensures that technological advancement serves broader social good while achieving business objectives.

Customer trust and regulatory compliance are inseparable elements of successful AI governance in insurance. Transparent AI practices build customer trust by ensuring fairness and accountability in all insurance interactions and decisions. Regulatory adherence reduces legal risks while ensuring compliance with rapidly evolving industry standards and government requirements. The combination of transparency and compliance creates a powerful foundation for customer confidence and loyalty, supporting long-term business success in an increasingly regulated environment.

Robust AI auditing practices ensure ongoing model reliability and compliance with ethical standards. Model accuracy assessment through regular auditing verifies that AI systems maintain reliable performance and produce correct outcomes over time. Fairness evaluation prevents bias development and ensures equitable treatment across all user groups and demographic segments. Compliance monitoring ensures AI systems continue meeting legal and ethical standards as regulations evolve. These comprehensive auditing practices build user trust while protecting organizations from operational and regulatory risks.

Continuous monitoring is essential for maintaining AI model performance in dynamic insurance environments. Ongoing performance surveillance tracks model outputs to identify any decline in accuracy or effectiveness over time. Detection of model drift identifies changes in input data or environmental conditions that cause predictions to become less accurate. Timely intervention capabilities allow for prompt model updates or retraining to maintain accuracy and reliability. This proactive approach prevents performance degradation and ensures consistent value delivery from AI investments.

Responding effectively to regulatory and operational challenges requires proactive management approaches. Anticipating regulatory changes helps maintain continuous compliance and reduces the risk of violations as standards evolve. Timely identification and resolution of operational challenges leads to improved efficiency and operational excellence. This forward-thinking approach positions organizations to adapt quickly to changing requirements while maintaining high performance standards. Proactive challenge management supports sustainable AI deployment that can evolve with changing business and regulatory environments.

Strategic AI governance in insurance requires integrating four essential elements for sustainable success. Building strong AI capabilities provides the foundation for effective governance through skilled teams and robust processes. Risk management ensures AI systems operate safely and reliably within complex insurance environments. Ethics promotion builds trust and upholds responsible business practices that serve both customers and society. Vigilant auditing maintains compliance and transparency in all AI applications. Together, these elements create a comprehensive framework for responsible AI deployment that drives business value while maintaining the highest standards of ethical operation.

Chapter 13: Health and Life Insurance

AI-powered health monitoring represents a fundamental shift from reactive to proactive healthcare management within insurance frameworks. Through continuous monitoring via wearable devices and smart sensors, AI systems collect real-time health data that provides unprecedented insights into individual health status. The technology's ability to identify potential health issues before symptoms manifest enables early intervention strategies that significantly improve patient outcomes. This proactive approach not only enhances policyholder wellbeing but also reduces long-term costs for insurers by preventing more serious health complications that would require extensive treatment and larger claims payouts.

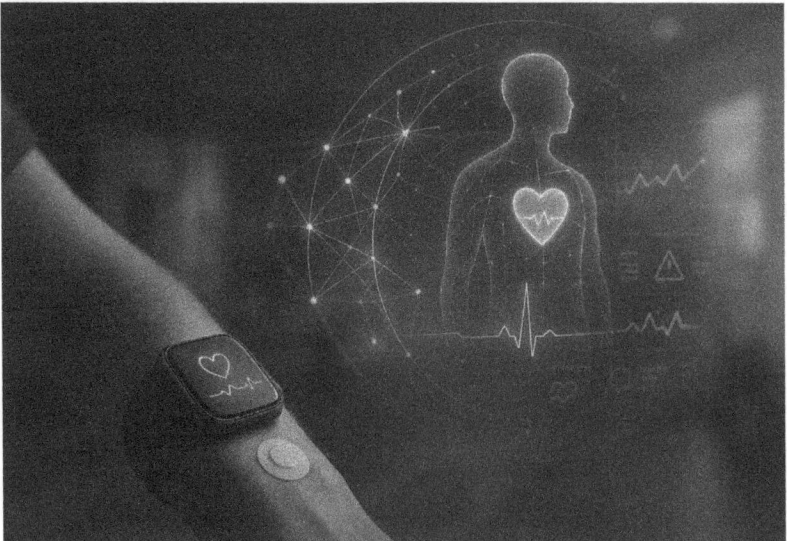

Personalized wellness programs powered by AI algorithms represent the next evolution in preventative healthcare within insurance. These systems analyze individual health data, lifestyle factors, and personal preferences to create tailored wellness plans that resonate with each policyholder. Insurers can provide customized coaching support that addresses specific risk factors and behavioral patterns unique to each individual. The personalization aspect significantly improves engagement rates and adherence to wellness programs. By helping policyholders reduce their health risks through targeted interventions, these AI-driven programs create a win-win scenario where individuals achieve better health outcomes while insurers benefit from reduced claims frequency and severity.

Predictive analytics for disease prevention leverages the power of big data and machine learning to identify health risks before they become clinical realities. These sophisticated models process vast datasets from multiple sources to uncover subtle patterns and correlations that indicate potential health risks. By forecasting disease onset with remarkable accuracy, AI enables both insurers and policyholders to implement preventive measures proactively. This predictive capability transforms traditional insurance models from simple risk transfer mechanisms to active health management partnerships. The insights generated support evidence-based decision-making for both individual health strategies and population health initiatives, ultimately reducing the burden of disease on both personal and societal levels.

The utilization of big data for individualized risk assessment represents a paradigm shift in insurance underwriting precision. AI systems integrate diverse data sources including medical records, lifestyle information, genetic factors, and environmental variables to create comprehensive individual health risk profiles. This enhanced data integration enables underwriting processes to become significantly more precise and reliable than traditional methods. The result is customized insurance products that are tailored to individual risk profiles and specific needs. This personalization benefits both insurers through more accurate risk pricing and policyholders through more appropriate coverage options and potentially lower premiums for lower-risk individuals.

Dynamic risk scoring and adaptive policy pricing represent the evolution toward real-time insurance models that respond to changing individual circumstances. Unlike traditional static risk assessments, these AI-driven systems continuously update risk scores using the latest available data to accurately reflect evolving health profiles. This dynamic approach enables insurance pricing and coverage to be adjusted in real-time, aligning with updated risk assessments. Policyholders benefit from pricing that more accurately reflects their current risk status, while insurers achieve more precise risk management. This creates a more equitable and responsive insurance ecosystem where premiums and coverage adapt to reflect actual rather than historical or projected risk levels.

Ethical considerations in personalized risk models are paramount to maintaining public trust and ensuring equitable insurance practices. AI-driven risk models must be designed and implemented to treat all customers fairly without introducing or perpetuating bias based on protected characteristics. Transparency in modeling approaches is essential so stakeholders can understand how risk assessments are calculated and decisions are justified. Data privacy protection remains critical, requiring robust security measures and clear consent processes.

These ethical frameworks ensure that while AI enhances insurance precision and efficiency, it does so in ways that maintain fairness, protect individual rights, and comply with regulatory requirements across different jurisdictions.

Incorporating wearable device data into insurance workflows creates unprecedented opportunities for continuous health monitoring and lifestyle-based underwriting. Wearable devices provide continuous streams of data including activity levels, sleep patterns, heart rate variability, and other vital signs that offer real-time insights into individual health status. This data integration allows insurers to assess lifestyle risks with greater accuracy than traditional methods. Furthermore, insurers can implement reward systems that recognize and incentivize healthy behaviors demonstrated through wearable data. This creates positive feedback loops where policyholders are motivated to maintain healthier lifestyles, resulting in improved health outcomes and reduced insurance claims over time.

Behavior-based underwriting and dynamic policy adjustment represent sophisticated applications of AI in analyzing human behavior patterns for more accurate risk assessment. AI algorithms examine behavioral data to identify patterns that correlate with health risks, enabling more nuanced and precise underwriting decisions. Dynamic policy adjustment capabilities allow insurance policies to respond in real-time to behavioral changes, creating more responsive and personalized insurance products. These systems actively encourage positive lifestyle changes by rewarding healthy behaviors with premium adjustments or additional benefits. This approach transforms insurance from a passive risk transfer mechanism into an active partner in promoting and maintaining policyholder health and wellbeing.

Data privacy and customer acceptance challenges represent significant hurdles that must be addressed for successful implementation of wearable-based insurance programs. Personal wearable data raises substantial privacy concerns that require careful management through robust security protocols and transparent data usage policies. Building and maintaining customer trust is essential for widespread adoption, requiring clear communication about how data is collected, used, and protected. Insurers must implement comprehensive data protection policies and obtain explicit, informed consent from users. Success depends on demonstrating clear value propositions to customers while maintaining the highest standards of data security and privacy protection to encourage voluntary participation in these innovative programs.

Automating claims processing with AI algorithms revolutionizes traditional manual processes by introducing unprecedented speed, accuracy, and efficiency. AI streamlines claim verification through automated validation of information, significantly reducing manual errors and processing times. Advanced fraud detection algorithms analyze patterns and anomalies across claims data to identify potentially fraudulent submissions with greater accuracy than human reviewers. Automated documentation review capabilities enable faster claim settlements while reducing administrative costs. These improvements benefit both insurers through operational efficiencies and policyholders through faster claim resolutions and improved customer service experiences. The technology transforms claims processing from a burden into a competitive advantage.

Enhancing accuracy and speed in disability assessments through AI addresses traditional challenges in subjective evaluation processes. Machine learning models provide objective evaluation of disability claims by analyzing comprehensive medical records and claimant histories using standardized criteria. This automated approach ensures consistent and fair decisions across all disability claims, significantly reducing human bias and variability in assessments. The enhanced accuracy and speed of AI-driven assessments improve customer satisfaction through faster claim resolutions and more predictable outcomes. This technological advancement creates more equitable processes while reducing administrative burdens and improving overall efficiency in disability insurance operations.

Supporting long-term care and chronic illness management through AI creates comprehensive systems for ongoing health monitoring and care optimization. AI tools continuously monitor patient health status to detect changes and facilitate timely interventions for chronic conditions. Predictive capabilities enable AI to forecast future care requirements, ensuring appropriate resources and support are available when needed. Personalized treatment plans based on individual patient data optimize care delivery and improve health outcomes. This proactive approach to chronic illness management benefits policyholders through better health outcomes and insurers through more effective cost management and reduced long-term care expenses.

In conclusion, AI transformation in health and life insurance delivers comprehensive benefits across all operational areas. Enhanced wellness programs leverage AI to analyze health data and provide personalized recommendations that improve policyholder engagement and health outcomes. Personalized risk assessments integrate multiple data sources for unprecedented accuracy in insurance evaluations, benefiting both insurers and policyholders through more appropriate pricing and coverage. Improved claims processing increases operational efficiency while reducing errors, creating better experiences for all stakeholders. These AI-driven innovations represent the future of insurance, creating more responsive, accurate, and customer-centric insurance products and services that benefit the entire industry ecosystem.

Chapter 14: Property and Casualty Insurance

Property and casualty insurance forms the backbone of risk management for individuals and businesses. This insurance category encompasses two primary components: property insurance, which protects against physical damage or destruction of assets like homes, buildings, and personal belongings, and casualty insurance, which provides liability protection when policyholders are legally responsible for injuries or damages to others. The scope includes essential coverages such as homeowners insurance, auto insurance, and comprehensive commercial policies that protect businesses from various operational risks and liabilities.

Property and casualty insurance addresses four fundamental risk categories that affect both personal and commercial policyholders. Fire coverage protects against one of the most devastating and common property risks, safeguarding against structural damage and content loss. Theft protection ensures compensation for stolen valuables and property. Natural disaster coverage becomes increasingly critical as climate change intensifies weather-related risks, covering floods, hurricanes, earthquakes, and storms. Liability and accident coverage protects policyholders from financial devastation when they're found legally responsible for causing injury or property damage to third parties.

The importance of property and casualty insurance extends far beyond individual protection to support broader economic stability. For individuals, insurance provides crucial asset protection, preventing catastrophic financial losses that could destroy personal wealth and security. Businesses rely on comprehensive coverage to manage operational risks and ensure business continuity during challenging circumstances. On a macro level, insurance supports economic resilience by distributing risk across large pools of policyholders, preventing individual catastrophes from creating widespread economic disruption and enabling faster recovery from major disasters and incidents.

Geospatial analytics revolutionizes insurance risk assessment by leveraging sophisticated geographic data analysis. This technology enables insurers to identify and quantify location-specific risk factors with unprecedented accuracy, moving beyond broad generalizations to precise, data-driven insights. Geographic data analysis incorporates multiple variables including terrain, weather patterns, proximity to water bodies, and historical claim data. The identification of flood zones and earthquake-prone areas allows insurers to make informed decisions about coverage availability and pricing, ensuring that policies accurately reflect the true risk exposure of specific locations.

Advanced mapping techniques utilize sophisticated software and data visualization tools to identify and categorize risk-prone areas with remarkable precision. These systems process vast amounts of geographic, meteorological, and historical data to create detailed risk maps. Risk categorization involves classifying regions into different risk levels based on comprehensive data analysis, enabling insurers to develop appropriate coverage strategies for each category. This approach facilitates tailored insurance coverage, allowing companies to customize policies based on specific geographic risks while prioritizing mitigation efforts and resources for the highest-risk areas.

Real-world applications of geospatial analytics demonstrate significant improvements in insurance operations. The technology enhances the accuracy of predicting potential insurance losses by incorporating location-specific variables that traditional models often overlook. Geospatial tools enable insurers to optimize underwriting decisions by providing detailed risk assessments for different geographic areas, leading to more precise pricing and coverage decisions. Market-specific applications show how geospatial analytics can be tailored for various insurance markets and regions, improving decision-making processes and ultimately benefiting both insurers and policyholders through more accurate risk assessment.

Artificial intelligence is transforming disaster prediction through sophisticated data analysis capabilities. AI systems can process vast datasets from multiple sources including satellite imagery, weather stations, seismic monitors, and historical records to detect early warning signs of natural disasters with remarkable accuracy. These AI-powered early warning systems provide timely alerts that enable proactive disaster preparedness and response, potentially saving lives and reducing property damage. The technology supports comprehensive risk management strategies, helping insurers and emergency response teams minimize disaster impacts through better preparation and more effective resource allocation during critical events.

Machine learning integration represents a significant advancement in catastrophe modeling accuracy. These systems improve traditional models by processing diverse and complex data inputs, including non-linear relationships and patterns that human analysts might miss. The continuous learning capability allows models to update and enhance their predictions based on new events and data, creating increasingly accurate forecasting over time. This results in more reliable catastrophe risk estimates that support better decision-making for insurers, enabling more accurate pricing, appropriate reserves, and effective risk management strategies.

Real-time data integration creates dynamic forecasting capabilities that significantly enhance disaster preparedness. Continuous data streams from weather monitoring stations, satellite systems, and sensor networks provide vital information that feeds directly into AI forecasting models. These AI systems process incoming data dynamically, generating timely and precise forecasts that reflect rapidly changing conditions. The result is enhanced rapid response planning, where dynamic forecasts enable quick decision-making and effective emergency response. This capability is particularly valuable for insurers who need to prepare for potential claims surges and deploy resources efficiently.

Smart home devices are creating unprecedented opportunities for continuous data collection in residential properties. Common smart devices include sensors that monitor temperature, humidity, and air quality, sophisticated alarm systems that detect intrusions and emergencies, and intelligent thermostats that track usage patterns and environmental conditions. These devices enable continuous data collection on home conditions, occupancy patterns, and potential risk factors. The wealth of data collected provides valuable insights into potential risks, enabling proactive home security measures and predictive maintenance that can prevent losses before they occur.

Smart home data enables insurers to build comprehensive and accurate risk profiles based on actual behavior and environmental conditions rather than assumptions. Data analysis for risk profiling incorporates real behavioral and environmental information collected continuously from various smart devices throughout the home. This approach allows insurers to move beyond traditional demographic and location-based risk factors to include actual usage patterns, maintenance habits, and security behaviors. Behavior and condition identification through smart devices helps insurers predict and manage claim risks more effectively, leading to more accurate underwriting and personalized policy recommendations.

The integration of smart home data is revolutionizing underwriting accuracy and premium pricing strategies. Smart home data enhances underwriting by providing real-time information about home safety conditions, usage patterns, and risk behaviors that traditional methods cannot capture. Dynamic premium pricing adjusts rates based on data-driven risk assessments, encouraging policyholders to maintain safer homes through better security practices and preventive maintenance. This creates powerful incentives for safety, where policyholders may receive rewards, discounts, or lower premiums for demonstrating responsible home management through smart technology adoption and consistent safety behaviors.

Drones are revolutionizing property inspection through advanced aerial surveying capabilities. These unmanned vehicles perform comprehensive aerial property surveys, providing detailed views of rooftops, structures, and surrounding areas that are otherwise difficult or dangerous to inspect through traditional methods. Drones offer safe access to hazardous areas, eliminating the need to put human inspectors at risk when examining damaged structures, steep rooftops, or areas affected by natural disasters. This technology enables more thorough inspections while reducing safety concerns and inspection costs for insurance companies.

Drone imagery significantly improves claims processing efficiency and accuracy. The technology enables faster damage assessment by providing immediate aerial views of affected properties, dramatically reducing the time needed for initial damage evaluations. High-resolution drone images improve assessment accuracy by capturing detailed visual evidence that minimizes errors in damage evaluation and claims processing. This leads to streamlined claims processing overall, where faster and more accurate assessments result in more efficient claims handling, quicker settlements for policyholders, and reduced administrative costs for insurance companies.

Remote assessment capabilities using drones are particularly valuable during disaster response situations. Drones offer a fast and safe method to assess disaster damage from above, providing critical information about affected areas without putting assessment teams at risk in potentially dangerous conditions. This aerial assessment capability enables insurers and emergency response teams to plan more effective recovery efforts by understanding the scope and severity of damage quickly. The technology supports coordinated disaster response, helping teams prioritize resources, identify the most critically affected areas, and develop comprehensive recovery strategies based on accurate, real-time damage assessment data.

Chapter 14: Property and Casualty Insurance

The integration of these innovative technologies represents a fundamental transformation in property and casualty insurance operations. Advanced risk assessment through geospatial analytics and AI significantly improves the accuracy of risk evaluation, enabling insurers to make more informed decisions about coverage and pricing. Enhanced underwriting precision using smart home data and artificial intelligence allows for highly tailored insurance policies that reflect individual risk profiles more accurately than ever before. Efficient claims processing through drone imagery accelerates verification and assessment procedures, reducing processing time and costs while improving customer satisfaction through faster claim resolution and more accurate damage evaluation.

Chapter 15: Commercial and Cyber Insurance

Commercial insurance serves as a critical financial safeguard for businesses, protecting against operational risks that could otherwise result in devastating losses. Beyond mere financial protection, it ensures business continuity by enabling organizations to maintain operations despite unexpected disruptions, equipment failures, or liability claims. The importance of commercial insurance spans all industries, providing essential risk mitigation that allows businesses to operate with confidence. Without adequate commercial coverage, a single incident could force business closure, making this protection fundamental to economic stability. Modern commercial insurance must adapt to evolving business models and emerging risks.

Cyber insurance has evolved from a niche product to an essential business protection as digital threats multiply exponentially. It specifically addresses risks from data breaches, ransomware attacks, system compromises, and other cyber incidents that can cripple modern organizations. Beyond financial compensation for direct losses, cyber insurance provides crucial support for recovery costs, legal fees, regulatory fines, and reputation management. Additionally, insurers offer proactive risk management tools including security assessments, employee training, and incident response planning. This evolution reflects the reality that cyber threats now pose existential risks to businesses across all sectors.

Modern risk management faces unprecedented challenges as cyber threats evolve at breakneck speed, often outpacing traditional security measures and requiring constant adaptation of protection strategies. Regulatory environments have become increasingly complex, with organizations navigating multiple jurisdictions, compliance frameworks, and constantly changing requirements that can significantly impact operational risk profiles. Traditional static risk assessment tools prove inadequate in today's fast-paced technological landscape, where new vulnerabilities emerge daily and market conditions shift rapidly. These challenges demand dynamic, AI-powered solutions that can process vast amounts of data in real-time and adapt to emerging threats automatically.

AI revolutionizes cyber risk assessment through sophisticated machine learning algorithms that identify subtle patterns in cybersecurity data invisible to traditional analysis methods. These systems evaluate system vulnerabilities comprehensively, examining network configurations, software versions, patch levels, and security controls to determine potential attack surfaces. Real-time threat intelligence integration ensures risk scores reflect current threat landscapes, incorporating global threat feeds, vulnerability databases, and emerging attack techniques. Security posture analysis evaluates organizational resilience by examining security policies, employee training effectiveness, incident response capabilities, and overall cybersecurity maturity levels for comprehensive risk evaluation.

AI-driven risk scoring relies on diverse data sources to create comprehensive threat assessments. Network traffic analysis detects anomalies, unusual communication patterns, and potential indicators of compromise that might signal security breaches or reconnaissance activities. Historical incident records provide crucial context, enabling

AI to identify common attack vectors, seasonal threat patterns, and industry-specific vulnerabilities based on past security events. Current security configurations are continuously evaluated to assess system strength, identify misconfigurations, and detect potential weaknesses in defensive postures. External threat intelligence feeds provide real-time updates on emerging global threats, new vulnerability disclosures, and evolving attack methodologies.

AI-driven cyber risk scoring offers significant improvements in accuracy and efficiency, enabling rapid identification of threats and vulnerabilities that might take human analysts considerable time to detect. However, effectiveness heavily depends on data quality, as flawed or incomplete information can significantly compromise risk assessment accuracy and lead to poor decision-making. Model transparency and expert oversight remain critical requirements to ensure trustworthy assessments and maintain confidence in AI-generated risk scores. The constantly evolving nature of cyber threats presents ongoing challenges, requiring continuous model updates, retraining with new data, and adaptation to emerging attack techniques to maintain scoring effectiveness.

Behavioral analytics transforms fleet risk assessment by analyzing real driver behaviors including speed patterns, braking habits, acceleration profiles, and route choices to identify risky driving patterns that traditional methods might miss. This data-driven approach enables insurers to assess fleet risk far more accurately than demographic-based models, using actual driving performance rather than assumptions. The insights generated encourage safer driving practices through feedback systems, training programs, and performance incentives, creating a positive cycle that reduces accident rates and associated insurance costs while improving overall road safety for commercial fleets.

Chapter 15: Commercial and Cyber Insurance

Modern telematics devices and sensors provide continuous real-time data collection on vehicle usage, driver behavior, environmental conditions, and mechanical performance, creating comprehensive operational pictures. This constant monitoring enables immediate detection of potential risks, from dangerous driving behaviors to mechanical issues that could lead to accidents or breakdowns. AI analytics process this vast data stream to generate precise risk profiles for individual drivers, vehicles, and entire fleets, enabling more accurate underwriting decisions and personalized insurance products that reflect actual risk levels rather than broad generalizations.

Behavioral analytics fundamentally changes underwriting and premium setting by enabling behavior-based risk assessment that uses actual driving data rather than demographic assumptions to evaluate individual and fleet risks. This approach allows premium customization that directly reflects real risk levels, creating financial incentives for safer driving while ensuring fair pricing for responsible drivers. The result is improved loss ratios for insurers as better risk selection leads to fewer claims, reduced payouts, and more accurate pricing models that benefit both insurers and policyholders through more equitable premium structures.

Natural Language Processing enables AI systems to interpret complex policy language with remarkable accuracy and efficiency, parsing dense legal terminology, identifying key provisions, and extracting critical information from lengthy documents. The system automatically highlights important terms within policy documents, making review processes faster and more thorough while reducing the likelihood of overlooking crucial provisions. AI-powered issue detection flags potential problems, conflicts, or gaps in coverage, assisting both underwriters and customers in making informed decisions about policy terms, coverage adequacy, and risk management strategies before policies are finalized.

AI excels at identifying coverage gaps by detecting missing or insufficient insurance protection that could expose organizations to significant financial risk during claims events. The technology spots exclusion clauses that might limit claim eligibility, increase policyholder risk exposure, or create unexpected coverage limitations that could prove costly. By detecting these inconsistencies and potential issues early in the policy development process, AI assists both insurers and insureds in proactively addressing risks, improving coverage adequacy, and reducing the likelihood of disputes or coverage surprises when claims occur.

Enhanced contract analysis through AI improves policy transparency by enabling clearer communication with clients, reducing misunderstandings about coverage terms, and providing plain-language explanations of complex provisions. This technology supports regulatory compliance by helping insurers meet legal requirements, maintain consistency with regulatory standards, and ensure policies comply with evolving insurance regulations across different jurisdictions. The result is improved customer relationships, reduced regulatory risk, and more efficient policy administration that benefits all stakeholders in the insurance process.

AI leverages predictive analytics to forecast individual risk factors with unprecedented accuracy, analyzing vast datasets to identify patterns and correlations that human underwriters might miss. Customer data analysis enables the design of insurance policies specifically tailored to unique individual preferences, risk profiles, business characteristics, and coverage needs. This personalization goes beyond traditional demographic segmentation to consider actual behavior, specific circumstances, and individual risk factors. The result is more relevant insurance products that provide better value for customers while enabling insurers to price risks more accurately and competitively.

Personalized product offerings significantly increase customer satisfaction by providing coverage that directly addresses individual needs rather than one-size-fits-all solutions. Transparent pricing enhances customer trust by clearly explaining how premiums are calculated, what factors influence costs, and how customers can potentially reduce their insurance expenses through risk reduction activities. This combination of personalization and transparency builds stronger customer relationships, increases loyalty, and creates competitive advantages for insurers who can demonstrate clear value propositions to their clients through customized solutions.

AI-based customization faces significant challenges in data privacy protection, requiring robust safeguards to protect sensitive personal and business information used in risk assessment and policy customization. Bias in AI models presents fairness concerns, as algorithms may inadvertently discriminate against certain groups or perpetuate existing inequalities in insurance access and pricing. Ensuring transparency in AI decision-making and maintaining fair access to insurance coverage represents critical ethical imperatives. Insurers must balance the benefits of AI-driven personalization with responsible implementation that protects consumer rights, promotes fairness, and maintains public trust in insurance markets.

AI significantly enhances risk assessment accuracy and speed in both commercial and cyber insurance sectors, enabling more sophisticated analysis of complex risk factors than traditional methods allow. Policy management innovation through artificial intelligence streamlines operations, reduces administrative costs, and enables highly customized insurance solutions that meet specific customer needs. The ultimate result is improved customer satisfaction through AI-driven services that provide personalized, responsive solutions tailored to individual requirements. This transformation represents a fundamental shift toward more efficient, accurate, and customer-centric insurance delivery that benefits insurers, customers, and the broader economy through better risk management.

www.ingramcontent.com/pod-product-compliance
Lightning Source LLC
Chambersburg PA
CBHW060045210326
41520CB00009B/1272